T0231325

Lectures in Geochemistry

One has to master the explicit
before exploring the implicit

Lectures in Geochemistry

Aleksey B. Ptitsyn

Novosibirsk State University

CRC Press
Taylor & Francis Group
Boca Raton   London   New York   Leiden

CRC Press is an imprint of the
Taylor & Francis Group, an **informa** business

A BALKEMA BOOK

CRC Press/Balkema is an imprint of the Taylor & Francis Group, an informa business

© 2018 Taylor & Francis Group, London, UK

Typeset by Apex CoVantage, LLC

Library of Congress Cataloging-in-Publication Data

Applied for

Published by: CRC Press/Balkema
        Schipholweg 107c, 2316 XC Leiden, The Netherlands
        e-mail: Pub.NL@taylorandfrancis.com
        www.crcpress.com – www.taylorandfrancis.com

ISBN: 978-1-138-32525-8 (Hbk)
ISBN: 978-0-429-45051-8 (eBook)

Contents

Preface

This short course of lectures is designed for chemical, biological, and environmental audiences who, though lacking in geology basics, are quite experienced in chemistry and biology. The structure and contents of these lectures are aimed at giving the reader a general idea of the Earth's chemical life and its related global geological events as fully as allowed by the present-day concepts of natural history (Dubnischeva, 1997).

This course is based on lectures that for more than a decade were given by the author at the Department of Environmental Chemistry of Novosibirsk State University. It is not a tutorial in modern geochemistry; therefore, many traditional aspects of this science are omitted here. Nor is it a monograph, but rather a short record of lectures where repetition, a typical feature of lectures, is hardly avoidable. The intentionally chosen concise mode of presentation allows the reader to grasp the material in its very essence without unnecessary details.

The term *geochemistry* was formulated in 1838 by Swiss chemist Christian Friedrich Schönbein and means "chemistry of Earth," thus linking together chemistry and geology. Classical courses of geochemistry, based on the works by Vernadsky, Fersman, and other outstanding geochemists, are presented to university students at geological departments. Meanwhile, some scientists (e.g., Ugai, 2002) consider geochemistry as one of modern branches of chemistry that deals with extremely complicated multiparameter systems. In these systems, unexpected fluctuations and events "confusing" commonly recognized chemical laws may arise due to interaction of system components over a wide range of physical and chemical conditions.

Taking into account the increasing public interest in environmental problems, special attention is paid to environmental geochemistry and to the habitat of mankind (i.e., our headstrong and capricious biosphere), not only from geochemical, but also from biological viewpoints.

The contemporary world (Hawking, 2015) in which we live is a result of protracted evolutionary transformations of the biosphere and Earth as a whole. This is why, when considering geochemical events, one cannot but take into account time-caused changes. However, some prominent scientists believe that the fundamental problems of evolution are paid insufficient attention. Similar to many other things on Earth, evolution of complex systems, biologically inert (bioinert) ones in particular, obeys the laws of thermodynamics and synergetics and deserves a separate discussion.

Vernadsky credited living matter, that is, the totality of all living organisms on Earth, as a potent geochemical factor. His fundamental theories laid the groundwork for biogeochemistry that studies chemical processes in bioinert systems. Accordingly, the evolution of living systems must be considered in the context of the development of their geological "home." Aristotle (384–322 BCE) believed that nature cannot be divided into the organic and inorganic. Also, he put greater importance on "function" rather than on morphological features. According to Aristotle, the crucial functions of living organisms are self-preservation and reproduction. We read of similar ideas, but on a modern scientific basis, in Vernadsky's works.

Because the Earth's interior cannot be observed directly, the geological data represent a set of more or less reliable hypotheses underlain by analysis-accessible facts and various calculations. Although our knowledge of the Earth's inner structure and events is most obscure, Earth as a material body must obey the general laws of natural science, and hence one can discuss the trends of the Earth's development set by these laws. It is these trends, dependent on internal and external factors, that will be the focus of the current course of lectures, which hopefully will stimulate the reader's desire to participate in developing a system of laws governing deterministic geomodels capable of predicting the future of the biosphere, and hence of mankind.

Like any researcher, the author has some personal predilections, so he hopes to draw the reader's attention to some specific and less studied issues of modern geochemistry, such as the geochemistry of cryogenesis (Taysaev, 2007), the geochemistry of disperse systems (Starostin, Lebedev, 2014), the geochemistry of microorganisms (Zavarzin, 2003a), and the geochemistry of the crypto biosphere.

Lecture 1

Geochemistry tools,
statistical geochemistry,
equilibrium thermodynamics: phase diagrams,
nonequilibrium thermodynamics,
thermodynamic principles of evolution,
laws of self-development

The field of geochemistry studies the distribution and amounts of chemical elements and their behavior on Earth and on the related planets. According to Vernadsky, geochemistry deals with geological processes at the "atomic level" and the history of atoms in the Earth's crust and on the planet as a whole. An alternative definition given by Tauson states that geochemistry is a science that studies the chemistry of geological events and the laws of migration, concentration, and scattering of element atoms in endo- and exogenous environments that manifest themselves on Earth.

Methodologically, geochemistry is traditionally divided into the geochemistry of elements, the geochemistry of processes, and the geochemistry of systems. Geochemistry of elements studies and describes the behavior of elements in conditions of various natural or human-caused systems and processes. It is focused on internal migration factors, such as the properties of chemical elements resulting from the structure of their atoms, namely, electronegativity, valence, the preferred mode of bonding in compounds, etc. Geochemistry of processes considers the behavior of various elements during specific endo- or exogenous processes. Lastly, geochemistry of systems deals with the behavior of some geochemical ensembles united by direct and reverse links. The methodological basis used here includes the theory of systems, laws, and regulations of thermodynamics and synergetics.

In combination with other sciences, geochemistry contributes to our knowledge of Earth as an extremely complex geosystem, including its origin, structure, evolution, and interactions between its constituents (Solov'eva, 2013). Like any science studying the material world, geochemistry is interdisciplinary: it comprises elements of physics, chemistry, and biology and uses tools developed in mathematics, informatics, thermodynamics, the theory of systems, etc. Nature is unity; it cannot be divided into individual parts considered separately from one another.

Only the Earth's surface and relatively minor depths are accessible for direct observation. The world's deepest bore hole, on the Kola Peninsula, barely reached a depth of 12 km, whereas the Earth's radius is more than 6,000 km. This is why our ideas concerning the inner parts of Earth are only hypotheses based on the results of some remote sensing techniques (mostly seismic sounding), manifestations of geological events on the Earth's surface, and various calculations and models. This makes it clear why the global geological paradigm, in part or as a whole, periodically undergoes alteration.

Seismic sounding

Using explosions or special powerful vibrators, seismologists transmit seismic waves into Earth from where they are partially reflected back at the interface between geospheres. With the known (at least approximately) velocity of seismic wave propagation in the rocks, it is possible to roughly estimate (with unavoidable errors, alas) the depth of this interface.

The seismic waves can be either longitudinal or transverse. The longitudinal waves are called P-waves (for "primary" waves), whereas the transverse waves are called S-waves ("secondary" waves). P-waves transmit a change in volume (compression stretching), whereas S-waves transmit changes in the shape of the material (the vibrations are at right angles to the direction of travel). The transverse waves can only travel through solids, because liquids block their passage. The wave speed is dependent on the material density. Seismic sounding was used to outline the boundaries of the geospheres and to determine the Earth's macrostructure, but it cannot reveal the physical nature of these boundaries, such as their chemical and mineralogical composition or a changed phase state.

Understanding and description of so complex a system as Earth requires a sufficiently universal set of methods applicable to essentially different objects, including inert (inorganic), living, and even social (anthropogenic) ones. It is thermodynamics that satisfies to this requirement. In this course we will use methods of equilibrium thermodynamics, primarily the geometric thermodynamics (phase diagrams) convenient for perception (Day and Selbin, 1969; Garrels and Crist, 1968; Karapet'yants, 1949; Knorre *et al.*, 1990; Parmon, 1998; Ptitsyn, 2006; Smith, 1968). Also, we will use some rules of synergetics, generalized and formulated by Glensdorf and Prigogine, as a general criterion of evolution that allows for the appearance of dissipative structures and the oscillation regime in chemical reactions. A dissipative structure is an open system, a stable nonequilibrium state of matter created due to scattering (dissipation) of incoming energy. The stable state in Progogine's thermodynamics may be believed to be equivalent to the equilibrium state, according to Gibbs' *Thermodynamics* (Glansdorf and Prigogine, 1973; Haase, 1967; Prigogine, 1960; Purtov, 2000).

Nonequilibrium thermodynamics developed by Prigogine and his colleagues allows us to leave the thermodynamics of equilibrium systems for that of stationary systems. Glensdorf and Prigogine's universal criterion of evolution is an indirect consequence of the second law of thermodynamics for nonequilibrium systems. In the stationary state, the free energy of a system is consumed most economically (i.e., the system's efficiency is maximal).

Application of the theory of stationary states to general problems of evolution was considered by its authors to be in principle unjustified, because the essence of evolution is not in achieving a stationary state, but in its continuous development and improvement.

The triad of the theory of evolution includes variability, heredity, and selection. Variability is stochasticity, uncertainty, probability; heredity is determinism, predetermination based on previous events; selection implies laws that govern the process (e.g., the second law of thermodynamics, laws of self-development, etc.).

Evolution representing an unlimited (because there's no limit to perfection) sequence of processes lies in a nonlinear region far from thermodynamic equilibrium. Such processes possibly occur only in systems that are capable of self-organizing and self-developing through a complex network of direct and inverse connections. Thus, the laws of self-development become the governing shaft of evolution.

In its early years, geochemistry used methods "borrowed" from chemistry, mineralogy, physics, and other sciences. Later, it demanded its own tools. Because geochemistry, as a rule, deals with multiparameter systems and processes, it is expedient to include in its theory mathematical analysis, statistics, and probability theory, and then the theory of information and cybernetics. The basics of statistical geochemistry are virtually nonexistent, though their necessity for theoretical geochemistry is evident. An appeal to develop them was made by Grigory Aronovich Bulkin in his book *Introduction to Statistical Geochemistry* (1972).

To elucidate dependences between structural/statistical and thermodynamic parameters in geochemistry actually means to analyze relationships between thermodynamic conditions favorable for the formation of geochemical systems on the one hand and the statistics of element distribution on the other. In its capacity as a system-state function, dispersion is a measure of scattering, and hence it must be connected with entropy, which is a measure of order and scattering (dissipation) of energy in thermodynamics.

The Earth is an open, evolving, self-developing system; the progressive development of which is determined by the presence of an external flow of energy, matter, and information. Continuous but impermanent in time, the energy impact on Earth from the cosmos, combined with internal factors, determines evolution of Earth as a planet and influences its internal processes. The multifactor nature of Earth as a system underlies numerous degrees of freedom; because of this, it behaves like a statistical ensemble and obeys the laws of self-development, which are the fundamentals of statistical geochemistry.

Let us consider these laws. Following Bulkin, the external energy coming into the system will be termed the basic reaction; changes that lead to a decrease in entropy of the system and its increasing complexity and diversity will be termed progressive changes, and by contrast, those contributing to system equilibrium and increasing entropy will be termed regressive changes.

The statistical law. With a sufficiently long reaction, the probability of a chain of successively progressive changes becomes higher than the probability of a chain of sequentially regressive changes.

The kinetic law. The most progressive path of development of a geochemical system is the most rapid ones; the fragments that follow these ways show a higher intensity of the basic reaction.

The energy law. Evolution of geochemical systems cannot but perform useful work in these systems. In the course of long transformations, the proportion of wasted energy decreases with the increasing share of energy consumed by useful work. A consequence of this law is the inevitable decrease in system entropy with increasing number of evolution stages.

The information law. This is the principle of maximal growth of evolutionary information. Each evolutionary change in the properties of an elementary site is imprinted as accordingly changed nature of the whole elementary fragment, and these changes are accumulated in the course of evolution.

The basic law of statistical geochemistry is as follows: the highest probability and speed are integral features of system self-development processes that lead to a maximum decrease in system entropy. System development is based on the mechanism of positive feedback when the result of the process strengthens it.

Bulkin's ideas were shared by other scientists. For example, Perelman (1989) believed that the continuous arrival of solar and subsurface energy into the Earth's crust determined its progressive development, thereby withdrawing its equilibrium state. The general law of the direction of evolution (minimal energy dissipation) may be formulated as follows: when a process can possibly develop in several directions allowed by the principles of thermodynamics, the direction providing the minimum energy dissipation (minimal entropy growth) is to be taken. Evolution is always aimed at reducing energy loss.

In the laws of self-development, the guiding effect of the second law of equilibrium thermodynamics is clearly manifested. Of all the possible paths of development, the system chooses the one that provides the most rapid and effective reduction of the total (its own plus incoming) internal energy. For this purpose, the efficiency of the energy supplied to the system is increased by spending a continuously growing part of it on useful work, including system complication, system structuring, and accumulation of information. Thus, the laws of equilibrium thermodynamics work even in progressively developing (far from equilibrium) systems.

Hence, the trend of evolutionary (progressive) development of Earth is predetermined by the laws of equilibrium and nonequilibrium thermodynamics, as well as the law of self-development. Importantly, when considering Earth's evolution, one should not oppose its geological evolution to biological evolution, because these are two branches of the same phenomenon. Of course you can study each branch separately, as Darwin did for living organisms, but this is just a specialization of the scientist.

As you know, at first Earth was flat. According to the views of ancient Indians, it rested on three elephants standing on a huge tortoise that swam in a vast ocean (it would be better to say – in the vast ocean of eternity, to link together space and time). At that prehistoric time, the most inquisitive people naturally wanted to look beyond the edge of Earth to see what was down there. That inspired the passion for travel to the edge of the Oecumene. The notion that Earth is a globe revolving similar to other planets around the Sun appeared as early as in ancient Greece. The new paradigm "turning" Earth into a globe at first caused a shock: it meant that on the other side of Earth people walked in the upside-down manner. It was the theory of relativity refuting the absoluteness of "up" and "down" that put everything in its place. Then Newton's theory of gravitation became useful, because it offered explanations as to why water does not drain off the Earth's surface and people remain there, too. Then Earth began rotating around its axis and turned into a geoid.

Later it was determined that Earth rotates around the Sun, being "tied" to it by the force of gravity. It should be noted that the physical nature of gravity is mysterious and has not yet been fully understood. The gravitational field and gravitational waves (lately much spoken and written about) are thought to spread almost as rapidly as light and never to be screened. Gravity is not a force acting in the passive background of space and time; according to Einstein's theory, it is a distortion of space-time itself (Dubnischeva, 1997). In other words, the gravitational field is the curvature of space-time. The gravitational interaction must be described by a quantum theory of gravity, which has not yet been developed. So, gravity is energy, and therefore it contributes to the overall energy balance of any system, including the system of planet Earth. The gravitational impact on Earth is determined by its position relative to other celestial bodies, which varies cyclically, because the cosmic orbits are elliptical. Consequently, some cyclic processes inside Earth may be associated with fluctuations in the intensity of gravitational energy in the near cosmos, rather than with some internal cyclic mechanism. This reveals a connection between geology (and geochemistry) and astronomy.

Lecture 2

Cosmochemistry.
The formation, composition, and structure of Earth

The patterns of element abundance in Earth together with the Earth's thermophysical, geological, and structural features were laid as early as at the stage of its formation about 4.5 billion years ago (Sharkov and Bogatikov, 2001; Uyeda, 1980). The age of the solar system is estimated at 5 billion years. Ancient philosophers believed that in the course of Earth's formation its rotation resulted in the separation of matter: heavy particles went to the center, whereas lighter ones remained at the periphery. Similar processes, in their opinion, took place in the cosmos. If so, the element distribution in the solar system must follow the same pattern as in Earth. Chemical characteristics of the planets of the solar system may depend on the sequence of their formation and their position relative to the Sun. According to present-day data (Dobretsov et al., 2001), the abundance of major elements in the terrestrial planets is as shown in Table 2.1.

Accordingly, the mechanism of Earth formation from cosmic matter determined its chemical composition and the primary distribution of chemical elements among the geospheres (i.e., the core, the mantle, and the crust). Recent data about the Earth's core show that it consists of iron, silicon, and nickel, whereas in its primitive (primary) mantle the sequence (in descending order) of the major elements is as follows: oxygen, magnesium, silicon, iron, calcium, aluminum. The composition of the Earth's crust is established

Table 2.1 The major chemical elements in the terrestrial planets

Planet	Average distance to the Sun (million km)	Major elements on the whole (in descending order)
Mercury	57.91	Fe, Ni
Venus	108.21	Fe, O, Si, Mg, Al
Earth	149.60	O, Fe, Si, Mg, Al
Mars	227.94	O, Si, Mg, Fe

with greater certainty. It comprises oxygen, silicon, aluminum, iron, calcium, etc.

The energy released during the Earth's formation from cosmic matter (upon accretion) warmed it up to fairly high temperatures. The average temperature gradient along its radius measured at accessible depths is 30°C per kilometer. As to temperature at greater depths, our notion is rather vague. A value of 3000°C at the boundary between the mantle and the core is believed to be consistent with the available data. The temperature of the Earth's core is estimated at about 6000°C.

Because the Earth was "molded" from cosmic matter, it is necessary to give a brief description of the latter using the following sources of information about the chemical composition of extraterrestrial material:

- Spectroscopic studies of the Sun, stars, interstellar space, comets, and atmospheres of other planets
- Identification of the nature of heavy particles present in cosmic rays and the solar wind using high-altitude balloons
- Analysis of compositions of meteorites
- Selenography data
- Theoretical calculations on nuclear fusion.

An approximate percentage of elements in the universe (Barabanov, 1985, with reference to Zeldovich) is as follows: H, 72%; He, 25%; C, O, Fe, Ar, N, Si, etc. in total, about 3%.

A lack of knowledge as to the properties of matter at ultrahigh pressures explains the uncertainty of our calculation-based notions about chemical compositions of the mantle and the core of Earth. The abundance of elements within Earth depends not only on the peculiarities of their formation (heavy elements are synthesized in the course of thermonuclear reactions in the nuclei of large stars, and the greater the atomic mass of an element, the more energy is needed for its formation), but also on stability of their nuclei. The most stable elements have even numbers of protons and neutrons. Elements with particularly stable nuclei are characterized by the so-called magic numbers of 2, 8, 20, 28, 50, 82, and 126, meaning the number of protons or neutrons. The series of magic nuclei looks like this:

$$_{22}He_4, \; _{2020}Ca_{40}, \; _{2832}Ni_{60}, \; _{3850}Sr_{88}, \; _{5070}Sn_{120}, \; _{5682}Ba_{138}, \; _{82126}Pb_{208}$$

Elements that have magical numbers of both protons and neutrons are called twice magical (He, O, Ca, Pb).

The Oddo-Harkins rule establishes the predominance of even elements over odd ones: the sums of clarkes of even and odd elements are in the ratio of 6:1. Neagley's law states that the degree of abundance of each element is a function of the stability of its nucleus.

Norwegian geochemist Victor Moritz Goldschmidt developed the basic geochemical law that the abundance of an element depends on stability of its atomic nucleus, whereas its distribution in nature is dependent on properties of the outer electron shell of its atoms.

At the beginning of the 20th century, Vernadsky came to a conclusion about the ubiquitous nature of all chemical elements; he said, "All elements are present everywhere."

The present-day structure of Earth and distribution of its intensive parameters (pressure and temperature) result from the mechanism of its formation and subsequent geological events (differentiation, evolutionary self-development, tectonics, etc.).

According to modern concepts (Sharkov and Bogatikov, 2001), the Earth's core sprang up first, followed by formation of the mantle around it; in other words, chemical elements were separated in the circumsolar space in the course of formation of the planets. The understanding of the early evolution of Earth came only with the results of Moon exploration available due to achievements in space science, because the tectonomagmatic characteristics of the Moon are close to those of Earth at the paleoproterozoic stage of its development (Sharkov and Bogatikov, 2001):

> At present we know practically nothing about geological processes that took place before the point of 4 billion years ago: the oldest rocks have exactly this dating. Since then, Earth has passed three stages of its development: (1) the nuclear stage that covered almost the entire Archaean and lasted to the point of 2.7–2.6 billion years ago; (2) the craton stage that belongs to early Paleoproterozoic from 2.6–2.5 to 2.0 billion years ago, and (3) the continental-oceanic stage which began in the late Paleoproterozoic about 2.0 billion years ago and continues to this day.
>
> (Sharkov and Bogatikov, 2001, p. 116)

Apparently, the liquid core of Earth arose about 2.6 billion years ago.

So, Earth consists of the core, the mantle, and the crust. Its average radius is 6300 km. The core comprises a liquid outer part, which blocks transverse seismic waves, and a solid inner part. The boundary between the core and the mantle is at a depth of about 2900 km. Within the mantle at a depth of about 1000 km there is a zone where the properties of matter change abruptly. That is why the mantle is divided into the upper and lower mantle (some scientists believe that the mantle has three layers, with the other interlayer boundary lying at a depth of about 550 km). The Earth's crust is a rather thin outer shell that is approximately 70–75 km thick under the continents and

10–15 km thick under the oceans. The boundary between the mantle and the crust, the so-called Moho, was named after Andrija Mohorovičić, the Croatian seismologist who discovered it. The Earth's crust is composed of two basic rock types: granite (the upper layer) and basalt (the lower layer). However, this does not mean that these layers really consist of basalts and granites. This means that seismic waves travel within them with the same speed as in basalts and granites. The boundary between the basalt and granite layers is called the Conrad boundary. Lately, many geologists have come to the conclusion that geosphere boundaries are manifested not by a change in the rock composition, but by phase transitions resulting from a change in the intensive parameters, that is, pressure and temperature. This is the only reasonable explanation of continuity of these boundaries and their location at a certain depth.

Additionally, we distinguish the Earth's asthenosphere and lithosphere; the former is a viscous, partially melted layer of the upper mantle, whereas the latter comprises the entire hard upper shell of the planet (above the asthenosphere), including groundwater and gases.

Different ideas have been offered about the transformation of Earth's surface. The latest, introduced Alfred Lotar Wegener's (1880–1930) mobilist theory, which proposes that before the Carboniferous period (about 300 million years ago) all modern continents were assembled together in a supercontinent, which he named Pangaea. Pangaea began breaking up, and its fragments moved apart. This idea was naturally inspired by the outlines of the continents.

At first Wegener's theory was generally criticized because it contrasted sharply with the contemporary paradigm. However, a follow-up study showed its validity, and it gradually developed (after 1950) into the concept of continental drift. Today, it is the leading concept associated with evolution of landscape geochemistry, the differentiation of biodiversity, and the pathways of human migrations. Isolation of continents from one another resulted in the conservation of endemic species (e.g., some primitive mammals currently absent from other parts of the world are still extant in Australia). However, some opponents of Wegener's theory believe that Pangaea never existed and that the continents have always been separate.

The average content of chemical elements in Earth (gross amount) was actually set during planet formation, and since then it has remained virtually unchanged (though this has not been proved). One can imagine two "ways" of changing the gross amount of specific chemical elements: (1) the exchange of matter with outer space (i.e., invasion of cosmic material into Earth with concurrent loss of terrestrial light elements, hardly very significant) and (2) radioactive decay (i.e., a decrease in the number of natural radionuclides with increasing amount of their decay products). It is noteworthy

Table 2.2 Abundance of chemical elements in the Earth's crust (Barabanov, 1985)

Content, $g\,t^{-1}$	Chemical elements
0.001–0.01	Re, Os, Ir, Ru, Rh, Te, Pt, He, Au, Pd
0.01–0.1	Ar, Se, Ag, Hg
0.1–1.0	Cd, Bi, In, I, Sb, Lu
1.0–10.0	Eu, Dy, Ho, Er, Yb, Hf, Ta, W, Tl, U, Ge, As, Br, Mo, Sn, Cs, Pr, Sm, Be
10.0–100.0	Pb, Th, Y, Nb, La, Ce, Nd, Li, B, N, Sc, V, Cr, Co, Ni, Cu, Zn, Ga
100.0–1000.0	C, F, P, S, Cl, Rb, Sr, Zr, Ba
1000.0–10,000.0	Mn, Ti
> 10,000.0	O, Si, Al, Fe, Ca, Mg, Na, K

that the decay products of natural radionuclides contain stable isotopes of some chemical elements. Thus, it is doubtful that the gross number of all stable elements remained unchanged during the Earth history.

The gross chemical composition of Earth determines, one might say, its multicomponent thermodynamic model. Moreover, because the amounts of different elements differ by many orders of magnitude (see Table 2.2), the roles of different elements in the planet's "life" are also dramatically different.

Geochemistry includes the concept of excess and scarce elements. Excess elements are those abundant for any reactions. In thermodynamic calculations, their chemical potential can be equated to unity. Deficiency of scarce elements prevents them from forming their own minerals, and therefore they enter other minerals in the form of isomorphic impurities. Thus, excess elements strike the keynote of geochemical processes, whereas scarce ones play the role of a "crowd." However, the boundary between these two types of elements demands a strict quantitative validation. In some cases the addition of a relatively small amount of some element can significantly change the properties (physical, thermodynamic, and others) of the system. A vivid example is the process of alloying in metallurgy.

In hydrogeochemical events, the role of the excess component is played by water, a special ("abnormal") compound with specific properties, which we will discuss separately.

Energy of Earth

All processes that occur in Earth are energy-consuming; the sources of energy deserve a special discussion, because they are associated with the driving forces of these processes.

The energy of geological processes in general and geochemical ones in particular is supplied from internal and external sources. The external sources of terrestrial energy are solar radiation (mostly), cosmic rays, and tidal friction.

In normal conditions the average amount of heat that Earth receives from the Sun is 34 KJ $(s \cdot cm^2)^{-1}$. The fate of this heat has been studied insufficiently. A part of the solar energy is reflected back into space, another part is absorbed by clouds, and certain portions are spent on warming up the Earth's surface and conservation of life (photosynthesis).

Tidal friction is apparently a less easily understood source of terrestrial heat. The fact is that, like the level of seas and oceans, solid matter of the continents undergoes tidal variations which, though insignificant in amplitude (up to 30 cm), contribute to rock warming up.

The gravitational field is a most important physical field that constantly influences geological bodies. The energy of a body with a mass m depends on its position in space. The change in the phase free energy of a mass M containing n_i moles of a component having the molecular weight M_i when it is raised (lowered) to a height h is equal to:

$$\left(\frac{\partial G}{\partial h} \right)_{P,T} = gM = g \sum_i n_i M_i \qquad \qquad 2.1$$

Given that the chemical potential of the component is defined as:

$$\mu_i = \left(\frac{\partial G}{\partial n_i} \right)_{P,T} \qquad \qquad 2.2$$

this can be written as:

$$\left(\frac{\partial \mu_i}{\partial h} \right)_{P,T} = gM_i \qquad \qquad 2.3$$

and can be used in calculating the element abundance in the Earth's crust.

Internal energy sources are residual heat of Earth and heat released during decay of natural radionuclides. In addition, energy is released in the course of mineralization (combustion) of organic residues. The major natural radionuclides are uranium, thorium, and ^{40}K. Because they are unevenly distributed in the rocks (their concentration in granitoids is about five times higher than in basaltoids), the "heat supply" of different areas is also uneven (Table 2.3).

Differences in "heat supply" of the areas should entail different heat fluxes on the Earth's surface, and hence local climate changes, differentiation of landscapes, and other important consequences.

Table 2.3 The heat flow on the Earth's surface (The Earth, 1974)

Types of areas		Mean value, $J/(s \cdot cm^2)$	Standard deviation
Dry land	Shields	3.9×10^{-6}	7×10^{-7}
	Orogenic	8.1×10^{-6}	2.1×10^{-6}
	Volcanic	9.1×10^{-6}	1.9×10^{-6}
Ocean	Basins	5.4×10^{-6}	2.2×10^{-6}
	Ridges	7.6×10^{-6}	6.5×10^{-6}
	Valleys	4.1×10^{-6}	2.6×10^{-6}

It has been calculated from thousands of measurements taken around the globe that the heat flux existing on the Earth's surface and leaving it for space cannot be driven solely by conductive heat transfer from deep horizons of Earth because the rate of heat transfer should be in agreement with the actual temperature gradient. In addition to conduction, heat can be carried by light. A hot body is shining, and the rocks are partially transparent to infrared and visible light. However, the light-induced increase in the heat transfer rate does not solve the problem either. It is convection in the mantle that appears to be the most efficient heat carrier. The possibility of convection is determined by the Rayleigh number (it must exceed 1500):

$$R = \frac{g\alpha\beta d^4}{hv} \qquad\qquad 2.4$$

In this equation, the thickness of the layer (*d*) is raised to the fourth power, which explains the possibility of convection in the mantle where viscosity (*v*) is very high and thickness is very large. Modern concepts of convection currents in the mantle are shown in Figure 2.1.

Thus, energy is transferred from the Earth's interior to the surface mainly due to convection-induced global circulation of hot mantle matter. This circulation also provides for cycles in global geological and geochemical processes that have been noticed to correlate with the galactic year (the period of rotation of the solar system around the center of the galaxy) of about 200 million years. Elliptical cosmic orbits cause cyclically changing distances between cosmic bodies, which result in cyclic fluctuations of gravitational energy coming to Earth; in turn, this entails periodic activation/attenuation of processes occurring in the Earth's interior.

Some people believe that a strong gravitational field can change the speed of time. No one calls into question the forward movement of time: it is undoubtedly irreversible. But its movement has never been proved to be linear – what if it moves along a Möbius strip? Well, this is from the realm

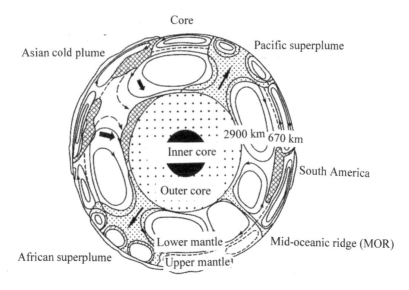

Figure 2.1 Scheme of convection currents in the Earth's mantle (Dobretsov *et al.*, 2001).

of fantasy of which the book *Monday Begins on Saturday* (1965) by the Strugatsky brothers is a perfect instance. Convection in the mantle is also associated with the drift of continents and the movement of slabs (plate tectonics).

For geochemistry dealing with plate tectonics, special zones characterized by active geochemical life are of special importance. These zones are as follows: mid-oceanic ridges known as oceanic spreading centers rich in volcanoes; subduction zones where the oceanic lithosphere (or, more often, the oceanic crust) of one plate slides beneath the continental lithosphere of another plate and sinks into the mantle (in this way oxidized rocks of the surface can increase the oxidation potential of the Earth's depths); and rift zones having a series of cracks that allow the intrusion to the surface of fresh portions of magmatic matter from the Earth's depths. Thus, tectonic processes are a global factor affecting physicochemical conditions in the Earth's interior.

It is important to mention that energy also circulates within Earth. The photosynthesis-generated energy accumulates deep in Earth in the form of chemically bonded organic compounds, from which it is released as heat during their decay. The share of this energy in the overall energy balance of Earth has not yet been calculated. It is not easy to make such a calculation because no one knows how much organic matter in the form of oil, gas, coal, etc. is buried in the Earth's depths.

Lecture 3

Geochemistry of elements

In geochemical events the role and behavior of chemical elements depend on both their properties and their amount. Under the same conditions different elements behave differently. Mendeleyev systematized the properties of elements in his periodic table of elements. In accordance with the laws of quantum chemistry, the properties of elements (chemical and physical) naturally vary with the charge of their nuclei. Table 3.1 shows Mendeleyev's periodic system of chemical elements as it looks today.

Table 3.1 Mendeleyev's periodic system of chemical elements

PERIODIC TABLE OF THE ELEMENTS

IA																	VIIIA
1 H	IIA											IIIA	IVA	VA	VIA	VIIA	2 He
3 Li	4 Be											5 B	6 C	7 N	8 O	9 F	10 Ne
11 Na	12 Mg	IIIB	IVB	VB	VIB	VIIB	—VIIIB—			IB	IIB	13 Al	14 Si	15 P	16 S	17 Cl	18 Ar
19 K	20 Ca	21 Sc	22 Ti	23 V	24 Cr	25 Mn	26 Fe	27 Co	28 Ni	29 Cu	30 Zn	31 Ga	32 Ge	33 As	34 Se	35 Br	36 Kr
37 Rb	38 Sr	39 Y	40 Zr	41 Nb	42 Mo	43 Tc	44 Ru	45 Rh	46 Pd	47 Ag	48 Cd	49 In	50 Sn	51 Sb	52 Te	53 I	54 Xe
55 Cs	56 Ba		72 Hf	73 Ta	74 W	75 Re	76 Os	77 Ir	78 Pt	79 Au	80 Hg	81 Tl	82 Pb	83 Bi	84 Po	85 At	86 Rn
87 Fr	88 Ra		104 Rf	105 Db	106 Sg	107 Bh	108 Hs	109 Mt	110 Ds	111 Rg	112 Cn	113 Uut	114 Uuq	115 Uup	116 Uuh	117 Uus	118 Uuo

Lanthanides series	57 La	58 Ce	59 Pr	60 Nd	61 Pm	62 Sm	63 Eu	64 Gd	65 Tb	66 Dy	67 Ho	68 Er	69 Tm	70 Yb	71 Lu
Actinides series	89 Ac	90 Th	91 Pa	92 U	93 Np	94 Pu	95 Am	96 Cm	97 Bk	98 Cf	99 Es	100 Fm	101 Md	102 No	103 Lr

However, discovery of new, superheavy elements has violated this periodic law, as evidenced by reports on properties of the element number 118 – oganesson (Og). The charge of its nucleus is so large that the orbit velocity of electrons must be increased in order to prevent their "sticking" to the nucleus. This element became the first "violator" of the periodic law, and although formally it falls into the group of inert gases, it does not look like its fellow groupmates.

Mendeleyev's periodic system was developed for normal pressure conditions when electron shells have a generally recognized form. It was suggested (Kapustinsky, 1956, etc.) that under high pressure some elements (starting with calcium, number 20) may undergo polymorphic transformations due to electron forcing in the unfilled 3d shell. For example, calcium can acquire the properties of bivalent titanium, and ferrous iron those of bivalent nickel. According to Kapustinsky (1956), in these conditions Mendeleyev's system turns from a seven-period system into a five-period one. Kapustinsky and his supporters call the Earth's mantle a "zone of degenerate chemism," and the core a "zone of zero chemism," where at ultrahigh pressure the atoms of elements cease to exist at all, and where there remain only nuclei "floating" in electron plasma. If these ideas are accepted by the majority of the world scientific community, the periodic system of elements will have one more coordinate – pressure. Then, at moderately high pressures, the periodic system of elements will gradually alter due to shuffled positions of the elements, and at superhigh pressures it will be transformed into a periodic system of nuclei (which was published in the late 1930s by Selinov, then a member of Kurchatov's team). Both of these systems are underlain by quantum mechanics and derived from the Schrödinger equation, thus being "brothers by birth."

Element phase transformation (if this term is applicable here) is another complication that must be taken into account when calculating the state of matter deep inside Earth and when considering chemical interactions in these extreme conditions.

However, in the Earth's interior, "among relatives," the behavior of elements also depends on geochemical regularities. In the theory of systems this is called the phenomenon of emergence, when a large system has properties that are not intrinsic to its constituent parts. Therefore, in addition to the canonical chemical classification (i.e., Mendeleyev's periodic system), several geochemical classifications demonstrate the behavior of elements in various geological situations (processes and systems). Among the most famous geochemical classifications are those offered by by Goldschmidt, Vernadsky, Fersman, Zavaritsky, and Berg. All of them are based on geological data about the behavior of elements in different geochemical processes and are useful for various geochemical studies as a tool for solving specific problems, but they cannot serve as a fundamental law that reveals the essence of things (however, we can consider them as precursors of such a law).

Geochemical classifications

The geochemical behavior of elements shows some specificity in comparison with its purely chemical prototype. This results from qualitative and quantitative differences in the composition of geosystems. In addition, note that in geochemistry we distinguish excess and scarce elements. From the standpoint of thermodynamics, it would be more rigorous to distinguish three categories: excess, system-forming, and scarce elements. Excess elements are those whose activity in reactions can be considered equal to unity. In other words, the system is open for such elements. System-forming elements are those that can be considered as thermodynamic components in the ratio that uniquely determines the figurative point of the system. Lastly, scarce elements are those characterized by a low content, which prevents them from forming their own phases and makes them join the phases of macroelements as isomorphic impurities. Accordingly, their geochemical behavior depends on the involvement of their host macroelements in the system.

Several geochemical classifications of elements have been developed. The simplest one divides elements into petrogenic and metallogenic elements. The group of petrogenic (rock-forming) elements includes oxygen, silicon, aluminum, calcium, magnesium, sodium, carbon, sulfur, etc. These are major components of the Earth's crust. Metallogenic elements form the productive part of mineral ore deposits. According to some researchers, metallogenic elements accumulate mostly during endogenous processes, whereas accumulation of petrogenic elements results from exogenous events. Obviously, this typification is very conditional and inaccurate. An illustrative example is iron with its uncertain position, which "sits on two chairs."

Vernadsky's classification of elements is based on their participation in chemical, and specifically, radiochemical processes, and distinguishes the groups of noble gases, noble metals, and of cyclic, dispersed, highly radioactive, and rare-earth elements. Inert gases are the products of radiochemical processes; they are chemically inert, and their geochemical role is far from being clear. Noble metals are also chemically resistant, due to which they change only slightly during geological processes. Cyclic elements constitute the bulk of the Earth's crust and Earth as a whole. According to the classification given first, they are excess and system-forming elements (silicon, aluminum, iron, oxygen, calcium, sodium, magnesium, sulfur, carbon, etc.). They create the main cycles of matter within Earth by passing from one state to another.

Concentrations of dispersed elements (Li, Sc, Ga, Br, J, Rb, Cs, In, Nb, Ta) are usually not high enough for the formation of independent minerals. So they join other mineral phases as isomorphic impurities and, according to Vernadsky, are components of "capillary" solutions.

Highly radioactive elements are, in fact, natural radionuclides (namely, Po, Rn, Ra, Ac, Th, U), whose recognition as a separate group is justified by their most important role in the energy of Earth.

Rare-earth elements having a special position in the periodic table display specific chemical and geochemical properties. Their minerals, which are, as a rule, endogenous in origin, appear to be most resistant under near-surface conditions.

In his classification of elements, Goldschmidt took into account their belonging to certain geochemical systems, affinity for sulfur and oxygen, structural peculiarities of ions, and magnetic properties, thus distinguishing the following groups:

- Atmophile elements that are essential constituents of the Earth's atmosphere
- Lithophile elements showing specificity to the silicate-oxide Earth's shell (crust, mantle)
- Chalcophile elements displaying a great affinity for sulfur
- Siderophile elements which are readily soluble in liquid iron
- Biophile elements that are the major constituents of living matter

This list could be rightly supplemented with hydrophile elements, major components of the hydrosphere, and technophile elements that are intensively used by industry. In 1954 Goldschmidt published his classification in the form of a table (Table 3.2).

It is easy to see that this classification is not rigorous (from the standpoint of the classification theory). First of all, the same elements may belong to different taxa (I show them in italic in Table 3.2). For example, oxygen is simultaneously present among atmophiles, lithophiles, and biophiles. Besides, depending on conditions, some elements can pass from one group to another (e.g., from lithophiles to chalcophiles). This is true for elements displaying amphoteric properties (Mo, W, As, etc.) and for transition elements (Mn, Fe, etc.). But of course there are "pure" representatives of a certain taxon (I show them in bold in Table 3.2). For example, inert gases belong exclusively to atmophiles; Na and Li are unambiguous lithophiles (yet, had hydrophiles been included in this classification, Na would have been among these, too); and Cu, Pb, and Zn are solely chalcophiles. Carbon is localized

Table 3.2 Geochemical classification of elements by Goldschmidt (1954)

Siderophile	Chalcophile	Lithophile	Atmophile	Biophile
Fe, *Ni*, *Co*, *P*, (As), *C*, **Ru**, **Rh**, *Pd*, **Os**, **Ir**, **Pt**, **Au**, **Ge**, **Sn**, **Mo**, (W), (Nb), *Ta*, (Se), (Te)	*S*, *Se*, **Te**, **As**, **Sb**, **Bi**, **Ga**, **In**, **Tl**, (Ge), (Sn), **Pb**, **Zn**, **Cd**, **Hg**, **Cu**, **Ag**, (Au), *Ni*, *Pd*, (Pt), *Co*, (Rh), (Ir), *Fe*, (Os)	*O*, (S), (H), **Si**, **Ti**, **Zr**, **Hf**, **Th**, **F**, *Cl*, *Br*, *I*, (Sn), **B**, **Al**, (Ga), *Se*, **Y**, **La**, **Ce**, **Pr**, **Nd**, **Sm**, **Eu**, **Gd**, **Tb**, **Dy**, **Ho**, **Er**, **Tm**, **Yb**, **Lu**, **Li**, **Na**, **K**, **Rb**, **Cs**, **Be**, **Mg**, **Ca**, **Sr**, **Ba**, (Fe), **V**, **Cr**, (Ni), (Co), **Nb**, *Ta*, **W**, **U**,	*H*, *N*, *C*, *O*, *Cl*, *Br*, *I*, **He**, **Ne**, **Ar**, **Kr**, **Xe**	*C*, *H*, *O*, *N*, *P*, *S*, *Cl*, *I*, (B), (Ca, Mg), (K, Na), (V, Mn, Fe, Cu)

Italic – elements in different groups
Bold – elements representing one group
Parentheses – elements that are not specific to this group

mostly to living organisms and their metabolic products, although according to Goldschmidt it also belongs to siderophiles and atmophiles. In general, this classification reflects only a kind of "addiction" of chemical elements to certain forms in which they exist in nature. However, despite the evident looseness of this classification, it is very useful, because it allows predicting mobility of a particular element in a certain geochemical situation.

The element's belonging to a specific geochemical category is determined by its ability to stably keep its phase for some time after it has been moved to a different environment in the course of a geological event.

Atmophiles are elements displaying the ability of remaining in the gas phase at low temperatures and pressures. These include inert gases, nitrogen, oxygen, hydrogen (in the form of H_2O), and carbon (in the form of CO_2). Mercury also displays some atmophilic features, and its total amount in the atmosphere is quite impressive.

When speaking about atmophiles, one cannot help mentioning sulfur, which easily passes into the gaseous phase in the form of SO_2 during volcanic eruptions and sulfide ore processing in metallurgy. As a result of technogenesis, the atmophilic properties are acquired (or increased) by some other elements, too.

Lithophiles are elements that in the course of Earth's matter differentiation concentrated in the silicate-oxide shell (crust, mantle). In the first place, these are silicon, aluminum, sodium, potassium and, to some extent, calcium (i.e., the major components of acid products of the differentiation). They are oxyphilic (i.e., prone to form oxygen compounds).

The group of lithophiles also includes lithium, boron, beryllium, and a number of other elements. Under specific conditions oxyphilic properties can be shown by classical chalcophiles. For example, this is true for Cu, Pb, and Zn in the hypergenesis zone where they can form both simple oxides (Cu_2O) and salts of oxygen acids.

Chalcophiles have a pronounced affinity for sulfur. They mostly form compounds with covalent bonding and belong to the side subgroups of groups I–V, including copper, silver, zinc, cadmium, mercury, lead, arsenic (III), antimony and bismuth. Chalcophilic properties can also be shown by bivalent tin and, depending on conditions, by the transition metals Mn, Fe, Co, and Ni. Also, sulfides produce other elements with medium strength characteristics (Mo, W, V, Cr), among which a sulfide compound is very common for molybdenum (MoS_2), while tungsten and chromium mainly form oxygen ones. Some rare-earth elements, such as Ge, In, and Ga, behave like chalcophiles, although they are usually present as isomorphic impurities in minerals of other more common elements.

The ability of metals to oxidize, and hence to exhibit their oxyphilic or chalcophilic properties, is related to the standard electrode potential, depending on which they are arranged in the following electromotive series:

Li, K, Ca, Na, Mg, Al, Mn, Zn, Cr, Fe, Co, Ni, Sn, Pb, H, Bi, Sb, As, Cu, Hg, Ag, Pt, Au

Those to the right of hydrogen are prone to exist in the native state. Among them, only noble metals are sufficiently resistant to chemical attacks to be able to remain in their native state even under hypergenic conditions. As a typical chalcophile, mercury is resistant to surrounding oxygen, but it readily undergoes oxidation in the presence of sulfide sulfur. Of native nonmetals, carbon and sulfur are widespread in nature.

Native carbon in the form of different types of coal, which is a good sorbent, is usually contaminated with various impurities.

Native sulfur, which is abundant in the areas of active volcanoes, is exceptionally pure. For example, some samples of sulfur taken from the Kurile Islands answer the brand "Superpurity." We even used it as a reagent in geochemical experiments.

One cannot but emphasize the great achievements of A. E. Fersman in the development of geochemical classifications (Fersman, 1959). Comparing the geochemical properties of chemical elements with their position in the expanded Mendeleyev's table (Table 3.3), A. E. Fersman revealed some

Table 3.3 Mendeleyev's table with expanded periods.

0	I	II	III	IV	V	VI	VII	VIII		I	II	III	IV	V	VI	VII	0
																H	He
He	Li	Be	B										C	N	O	F	Ne
Ne	Na	Mg	Al	Si	P								(Si)	(P)	S	Cl	Ar
Al	K	Ca	Sc	Ti	V	Cr	Mn	Fe Co Ni		Cu	Zn	Ga	Ge	As	Se	Br	Kr
Kr	Rb	Sr	Y	Zr	Nb	Mo	Ma	Ru Rh Pb		Ag	Cd	In	Sn	Sb	Te	J	Xe
Xe	Cs	Ba	TR	Hf	Ta	W	Re	Os Ir Pt		Au	Hg	Tl	Pb	Bi	Po	No 85	Rn
Rn	No 87	Ra	Ac	Th	Pa	U											
18	1	2	3	4	5	6	7	8	9 10	11	12	13	14	15	16	17	18

regularities in variation of clarks both within groups and within large periods of the table and discovered the diagonal rows of elements having close ionic radii, and therefore prone to isomorphism (Clark, 1961).

Fersman's classification explains the term "incompatible elements," whose behavior is somewhat specific. Incompatible elements are those not included in minerals of the upper mantle. The cause of their incompatibility lies in a large ionic radius or a high charge of the ion. The rare-earth elements (17 of them: scandium, yttrium, and the lanthanides) are a typical example of incompatible elements.

Fersman concludes that "in geochemistry, these [diagonal] rows suggest co-localization of these elements in nature" (Fersman, 1959, p. 469 [in Russian]). On the grounds of common chemical and geochemical properties, Fersman identified the following five families:

- Iron family: Ti, V, Cr, Mn, Fe, Co, Ni
- Molybdenum family: Y, Zr, Nb, Mo, Ru, Rh, Pd
- Rhenium family: Ta, W, Re, Os, Ir, Pt
- Uranium family: Rn, Ra, Ac, Tr, Pa, U
- Rare-earth elements

In Mendeleyev's table, Fersman singled out areas corresponding to certain geochemical associations, which are in good agreement with the above mentioned classification by Goldschmidt.

The large number of geochemical classifications is in itself a testament to their imperfection. Although all authors are guided by the periodic system of elements, each of them gives preference to a certain group of geological factors. As a result, for example, according to Goldschmidt, Vernadsky, and Berg, gold is grouped with noble metals, whereas according to Fersman and Zavaritsky it is a chalcophile. Unlike other authors, Berg classifies silver with noble metals, and not with chalcophiles. According to Goldschmidt and Berg, molybdenum belongs to the iron family, according to Zavaritsky it is one of the rare-earth elements, and according to Fersman it falls into the same group as platinoids and radioactive elements.

Differences, quite significant indeed, between classifications developed by these authors are caused by objective reasons, that is, by the multifactorial character of geochemical systems. To create a general geochemical classification of elements (geochemical law), some ingenious approach is apparently required. Meanwhile, there is a need for such a fundamental law. Mendeleyev's periodic system allowed for predicting and then discovering new chemical elements. Similarly, geochemists need such a tool in search of deposits of an unknown type. Thus far, geological exploration has used stochastic models based on known data and has been performed "by analogy." No object of an unknown type can be found in this way. It is necessary to develop a special theoretical basis of geochemistry.

Thus, the behavior of elements in geochemical processes is determined not only by their properties (internal factors), but also by their environment (external factors).

The mobility of elements during geochemical processes is largely determined by the forms in which they exist in nature; in turn, these forms depend on the type of chemical bonding characteristic of a given element. Also, the mobility is influenced by the ability of the element in question to demonstrate different valences in different physicochemical conditions. For example, under reduction conditions, tetravalent uranium is similar to thorium and chemically inert, whereas under oxidation conditions it is hexavalent and mobile, because as a uranyl-ion it exhibits the properties of alkali metals. Thallium, when monovalent, shows the properties of alkali metals, but when trivalent, it resembles chalcophiles and accumulates on the hydrogen sulfide geochemical barrier.

Geochemical classifications are based on geological empirics; that is, they group the elements into taxa basing on their correlation with known petrological or ore formations, on their ability to accumulate in the known types of ores, and on their belonging to known paragenetic associations.

Thus, the practical significance of these classifications consists in the possibility of predicting and searching for deposits of a known type. However, they allow for virtually no prediction of deposits of a new type, let alone search for these deposits. For this purpose, a classification based exclusively on the laws of nature without empirical subjectivism is required. As yet, few such laws exist in geochemistry, and those available usually lack a strict mathematical expression.

Thus, the development of a system of geochemical laws and rules is one of the most important problems. Let us mention some achievements in geochemistry that can serve as a basis for such a system. First, it is the empirical regularity revealed by Fersman (1959), which he called the basic law of geochemistry and formulated as follows:

> A succession of geochemical crystallization depends on the basic features of atoms (ions) present in the solution, melt or gas mixture, and usually follows a certain order according to which lattices of the densest packing, even symmetry, maximum valence and the smallest distances between their sites are the earliest to undergo crystallization; that is, on the grounds of modern energy concepts, these lattices have the maximum energy or, which is the same, the minimum reserves of free efficient energy; this order is consistent with Le Chatelier's principle and Ostwald's law, and in general, it follows the basic laws of thermodynamics, with the addition, however, that a special role in the succession is played not only by the amount of free energy, but also by its character expressed through the type of crystal structure (Kristallbauplan Niggli). (p. 464 [in Russian])

Next, it is the geogenetic law formulated by Rundqvist, according to which each smaller geochemical cycle reproduces the main features of a larger cycle. Additionally, there are laws of statistical geochemistry (Bulkin, 1972) that deal with the evolution of geochemical systems that have a flow of energy passing through them. All these empirical laws (better to say, regularities) still need a theoretical basis; this, by the way, was realized by Fersman (1959) himself, who wrote: "Surely my formulation can acquire an exact mathematical form only in the future" (p. 465 [in Russian]).

Force of the geochemical law(s) can be illustrated by the difference in behavior of rare-earth and dispersed elements. According to Vernadsky, dispersed elements are those never showing high concentrations in the Earth's crust (i.e., having weak concentration ability), although their amount can be significant (they may have relatively high clarks). These include, for example, Rb, Ga, Re, Cd, and Sc. In contrast, rare-earth elements, though less abundant in the Earth's crust, are able to form their own minerals and ore clusters. What is the reason for such a difference in behavior of these elements? It can hardly be explained only by properties of the elements themselves. Apparently, an important role is played by what we call external migration factors.

Thus, the existing geochemical classifications reflect the behavior of chemical elements either in a particular geochemical system (group of systems) or in specific geochemical processes. Methodologically, this explains division of geochemistry into geochemistry of elements, geochemistry of processes, and geochemistry of systems. The most valuable and useful classifications are those based not on separate empirical samplings (which vary in the process of cognition of nature), but on fundamental laws. Such a classification must have the most important ability of predicting the existence of yet unknown objects, like Mendeleyev's table that predicted yet undiscovered elements.

Logically, it can be proposed that the geochemical classification of elements should combine a classification of elements, a classification of geochemical processes, and a classification of geochemical systems. Presumably, then, it will be possible to take into account internal and external factors that affect element migration (i.e., geochemical behavior of elements) and to make predictions aimed at discovery of yet unknown geological situations.

Geochemistry of isotopes

At present, about 300 stable isotopes and more than 1200 radioactive ones are known (Ayvazyan, 1967; Urey, 1947; Walt, 1961, etc.). Some elements have one or two stable isotopes (they are members of odd groups in the periodic table), whereas others have more than two stable isotopes (all of them belong to even groups) (Table 3.4).

Isotopes of the same element have almost identical physical and chemical properties, but dramatically differ in nuclear ones. With the discovery in the 1930s of nucleon shells in atomic nuclei, the magic numbers N_m and Z_m, and also 2β-stable nuclides, it became possible to create a periodic system of atomic nuclei (Table 3.5). At the same time, a periodic system of isotopes was developed.

Table 3.4 Periodic table of elements showing the number of isotopes.

	1	*2*	*3*	*4*	*5*	*6*	*7*	*8*	
1	H-2								He-2
2	Li-2	Be-1	B-2	C-2	N-2	O-3	F-1		Ne-3
3	Na-1	Mg-3	Al-1	Si-3	P-1	S-4	Cl-2		Ar-3
4	K-2	Ca-6	Sc-1	Ti-5	V-1	Cr-4	Mn-1	Fe-4　Co-1	Ni-5
	Cu-2	Zn-5	Ga-2	Ge-5	As-1	Se-6	Br-2		Kr-6
5	Rb-2	Sr-4	Y-1	Zr-5	Nb-1	Mo-7	Tc	Rn-7　Rh-1	Pd-6
	Ag-2	Cd-8	In-2	Sn-10	Sb-2	Te-7	I-1		Xe-9
6	Cs-1	Ba-7	La-Ln-7	Hf	Ta	W-5	Re-2	Os-7　Ir-2	Pt-6
	Au-1	Hg-7	Tl-2	Pb-4	Bi-1	Po	At		Rn

Table 3.5 Periodic system of atomic nuclei according to Selinov (1990)

Element	Mass numbers
H	1.0081
D	2, 3
He	4, 5
Li	6, 7
Be	8, 9
B	10, 11
C	12, 13
N	14, 15
O	16, 17, 18
F	19
Ne	20, 21, 22
Na	23
Mg	24, 25, 26
Al	27
Si	28, 29, 30
P	31
S	32, 33, 34
Cl	35, 37
Ar	36, 38, 40
K	39
Ca	42, 43, 44
Sc	45
Ti	46, 47, 48, 49, 50
V	51
Cr	50, 52, 53, 54
Mn	55
Fe	54, 56, 57, 58
Co	57, 59
Ni	58, 60, 61, 62, 64
Cu	63, 65
Zn	64, 66, 67, 68, 70
Ga	69, 71
Ge	70, 72, 73, 74, 76
As	75
Se	74, 76, 77, 78, 80, 82
Br	79, 81
Kr	78, 80, 82, 83, 84, 86
Rb	85, 87
Sr	84, 86, 87, 88
Y	89
Zr	90, 91, 92, 94, 96
Nb	93
Mo	92, 94, 95, 96, 97, 98, 100, 102
Ma	?
Ru	96, 98, 99, 100, 101, 102, 104
Rh	103
Pd	102, 104, 105, 106, 108, 110
Ag	107, 109
Cd	110, 111, 112, 114, 116, 118
In	113, 115
Sn	112, 114, 115, 116, 117, 118, 119, 120, 122, 124
Sb	121, 123
Te	120, 122, 124, 125, 126, 128, 130
I	127
Xe	124, 126, 128, 129, 130, 131, 132, 134, 136
Cs	133
Ba	130, 132, 134, 135, 136, 137, 138
La	139
Ce	136, 138, 140, 142
59Pr	
Nd	
Sm	
Eu	
Gd	
Tb	
Dy	
Ho	
Er	
Tu	
Yb	
Lu	
Hf	
Ta	
W	182, 183, 184, 186, 188
Re	185, 187
Os	186, 187, 188, 189, 190, 192
Ir	191, 193
Pt	192, 194, 195, 196, 198
Au	197
Hg	196, 198, 199, 200, 201, 202, 204
Tl	203, 205
Pb	204, 206, 207, 208
Bi	209

Chemical elements in the "water–rock" system

In the lithosphere, the carrier of elements is water (aqueous solutions). Therefore, the ability of elements to pass from their solid phase to an aqueous solution is important. This happens in the system commonly known as the "water–rock system." Moreover, in geochemistry there is a separate investigation branch called "water–rock interaction." This includes a whole bunch of processes, namely, solid–liquid reactions (kinetics of reactions in heterogeneous systems), complexation, adsorption, the formation of nonautonomous phases on mineral surfaces, the effect of some elements on mobility of others, electrochemical processes, and even the vibrational modes of reactions and processes. All of them are well studied and included in courses of lectures on chemical kinetics, chemistry of complex compounds, electrochemistry, and others. Therefore, in the next lectures we will talk solely about the thermodynamics of nonautonomous phases, because this issue is still being discussed in the literature.

Chemical elements in bioinert systems

In bioinert systems, the behavior of elements follows a special pattern. Different organisms have different modes of adaptation to specific geochemical conditions (to different contents of elements), which underlies the selectivity in biological accumulation of chemical elements. For example, corn accumulates gold; tobacco leaves – lithium; pomegranate and nettle – iron; laminaria – iodine; death-cup – thallium and selenium; ascidium (a sea chordate) – vanadium; napier grass (which grows in Africa) – silicon dioxide; and so on. This is the source of biogenic mineral deposits.

Some organisms are able to develop mechanisms that protect them against "harmful" chemical elements. For example, for this purpose plants either use biochemical barriers of the rhizosphere (the root zone) or bind unwanted elements within their bodies for their subsequent removal with metabolic products. Recently, plants, insects, mollusks, and mammals have been reported to contain metallothioneins (low-molecular-weight proteins) that are capable of binding ions of heavy metals (i.e., zinc, cadmium, mercury, etc.). In addition, plants have a mechanism that protects their reproductive organs (i.e., flowers and fruits) from pollution. This is evidenced by distribution of heavy metals among plant organs: their highest concentration was detected in roots and leaves (needles), their lowest in flowers and fruits.

Uneven distribution of chemical elements on the Earth's surface leads to the formation of geochemical landscapes of a certain type, and hence to a peculiar chemical composition of the organisms living there; this causes endemic diseases (endemic osteoarthritis, goiter, caries, fluorosis, etc.). The unfavorable regions are those with either increased or decreased level of vital or toxic elements in the environment. And the deficit is sometimes worse than the excess.

Thus, when investigating the behavior of a particular chemical element, a geochemist must take into account a variety of factors, including geological, chemical, biological ones.

Lecture 4

The evolution of Earth as a planet

First of all, we need to determine the subject of our talk about evolution. We will not discuss biological evolution as it is seen by Charles Darwin, his supporters or opponents. Biologists have long been debating the theory of evolution; that is, whether it uses strong evidence or it is closer to faith than science, and whether such a theory can possibly exist at all. Opponents of this theory point out the absence of any intermediate links, which, in their opinion, is indicative of spasmodic rather than evolutionary character of development of living matter. Anyway, this is beyond the topic of our lectures. We will talk about evolution of the Earth system from the standpoint of thermodynamics and synergetics.

An indispensable condition for evolution of a self-developing system is its sufficient diversity, which provides the necessary degrees of freedom for a statistical choice (by trial and error) of the most favorable path of development. The diversity also ensures stability of the system in a troubled world of external and internal factors.

To the correct evolutionary path the system is guided by thermodynamics. In compliance with its laws, the system should minimize its entropy growth and choose a development option where the sum of its internal energy and energy coming from outside is minimal. For this purpose, the incoming energy should be consumed with the greatest possible efficiency. Thus, even in an evolving system that is very far from the equilibrium state, the second law of thermodynamics works, and the system, despite fluctuations, obeys the above mentioned laws of self-development. An important property of inorganic systems is their self-training ability based on experience that has been gained from previous stages. The larger the number of these stages, the more perfect self-training. Another important factor is emergence; that is, the system's coming into possession of properties that are lacked by its components. What exactly new property will emerge depends on the set of components and results from their interactions by a number of different mechanisms. For example, academician George Zavarzin defines life as an emergent property of a system of components united in one organism. Most important, when studying individual components of a system, it is virtually impossible to predict appearance of an emergent property. Therefore, the "bottom-up" study of natural systems (from simple to complex) has no future. To understand what causes the appearance of emergent properties, it is necessary to study interactions of the components.

In a large multicomponent system, which actually is the biosphere, probability theory principles play an important role. Why not assume that the appearance of emergent properties in such a system also obeys these principles? Then, a random combination of factors might cause some unexpected "outbursts" from the logical evolutionary sequence of Earth's species diversity.

The geological (prebiological) evolution of Earth began virtually from the moment of its formation (Sorohtin and Ushakov, 1991). There are no reliable material traces of the early Earth's history.

As one of the matter-differentiating mechanisms in the prime, still quite hot Earth with local, periodically arising melting sites, one can imagine something like zone melting, whereby lighter components are gradually forced upward by repeated rock melting. It is believed that the newly formed solid crust stopped melts coming from the mantle and made them spread horizontally, thus arranging for a kind of zone melting. However, this can only explain local differentiation of the Earth's matter, while the mechanisms of formation of continuous concentric geospheres must be quite different.

The Moon, regarded as an analog of the Earth's paleoproterozoic, is believed to have its primary crust formed as a result of plagioclase flotation in the global magma ocean (Ringwood, 1982).

Simultaneously with differentiation of material of the Earth's mantle there occurred partial degassing, thus originating the primary atmosphere, which presumably contained water vapor, methane, ammonia, carbon, and maybe hydrogen as well, and was of a reducing nature. Because the Earth's surface was still very hot, no hydrosphere could form.

When the surface of the planet cooled sufficiently, water vapor began to condense from the atmosphere and fill surface depressions, thus forming a proto-ocean, at first shallow. Later, the dismemberment of Earth's reliefs increased, and the oceans deepened. The hypothesis of isostasy (Clarence Dutton, Gerhard Pratt, and others) suggests that Earth tends to some isostatic equilibrium. Topographic heights in one place must be compensated for by depressions in another, for otherwise the Earth's rotation will fail. Therefore, the oppositely directed global processes like mountain formation/denudation (mountain demolition and removal of its material), transgression/regression of the seas, and others are not accidental but regular.

Nature never acts aimlessly. Although sometimes its actions seem retarded or too hasty, actually this is a normal "work" of impacts and feedbacks.

A most important event in the Earth's evolution is the appearance of life about 4 billion years ago. As to the origin of life, it is still an open question with three possible answers: (1) everything comes from God; (2) it came from the outer space; (3) it originated from inorganic material. In the latter case, at first there were only forerunners of life, that is, organic compounds (sugars and amino acids); it took them a very long time to gradually form the foundation of the future biosphere. The current concepts are that the primary living matter could only arise in an oxygen-free atmosphere, because oxygen was disastrous for primitive forms of life. Accordingly, the entire biochemistry of primary organisms was adapted to a reducing environment.

The first organisms, prokaryotes, appeared about 3.5 billion years ago. At an early stage, unicellular prokaryotes living on chemosynthesis and photosynthesis acted as primary producers. Little by little, they created a basis for the nitrogen-oxygen atmosphere. Later, via feedbacks prokaryotes began to control the atmosphere composition, which can be regarded as the origination of the biosphere. Here again it is appropriate to quote Zavarzin (2003b): Abiotic environment is of decisive importance for the existence

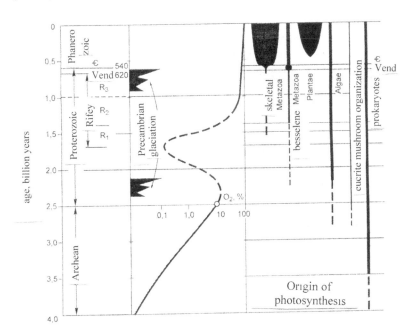

Figure 4.1 The major events in evolution of the Precambrian biosphere (Abyzov *et al.*, 2002).

of life. Therefore, the Earth's evolution must not be divided into geological and biological, although living organisms follow their own laws of evolution, for example, the law of evolution unevenness (uneven phylogenetic development), according to which the rate of evolution differs from taxon to taxon. The vital functions of primary organisms resulted in a gradually increasing content of oxygen in the Earth's atmosphere, thereby contributing to the next critical event – the advent of an oxygen atmosphere.

About 2.5 billion to 3.0 billion years ago the oxygen content in the atmosphere reached at first 1% of the present value (first Pasteur point), when aerobic organisms appeared, and then 10% (second Pasteur point), when life came to dry land. For some organisms, this was a stress, because they had to adapt their biochemistry to living in oxidizing conditions.

In the opinion of Zavarzin, it is cyanobacteria that are responsible for formation of the oxygen protoatmosphere; these are members of the persistent group (from Latin *persistere* meaning "to be continuing or permanent" or "to have continuity of phylogenetic characteristics") that exists from ancient times to the present day. Many microorganisms are resistant to hard ultraviolet radiation, and therefore photosynthesis could take place not only in the ocean under a layer of water (as many researchers still believe), but also in shallow waters on land. Traces of cyanobacterial mats (stromatolites) have been found in rocks that date from more than 2 billion years ago.

At present, cyanobacteria can be found in geologically active zones (Kamchatka, the Kuriles) and in soda lakes, which, according to Zavarzin, are relics of the protobiosphere. Because of their abundance, cyanobacteria have always been and remain active suppliers of oxygen to the atmosphere.

Gradually, the growing oxygen content reached a level that resulted in formation of an ozone layer in the upper atmosphere that shielded Earth from the hard ultraviolet radiation. Then, a myriad of living organisms poured onto land. That was the beginning of a new important stage in the Earth's evolution – an intensive growth of biodiversity and formation of the biosphere as a self-developing system. The biosphere became involved in building and proper maintenance of geochemical cycles. The increasing oxygen proportion in the atmosphere was accompanied by geochemical rearrangements in the lithosphere. According to some researchers, this testifies to destruction of continental rocks in oxidizing conditions and carrying weathering products to shallow waters. Intensive oxidation of crystal rocks on the Earth's surface caused the formation of sedimentary rocks with an elevated oxidative potential; this potential also increased in the Earth's interior due to subduction processes (lithospheric plate/oceanic crust sinking into the mantle).

The main problem concerning the similarity of laws governing abiocoen and animate nature is that in both cases evolution leads to the realization of unlikely scenarios. The evolution of both living and inanimate systems is antientropic (Milne *et al.*, 1985).

The next most important landmark in the Earth's evolution is the appearance of humans, who are the only biological species that by their own will (or whim) can determine the pathway of their development, thus affecting patterns of nature's evolution. The nearly million-year long history of mankind has demonstrated what humans are capable of. According to Vernadsky, by the beginning of the 20th century, the human impact (technogenesis) had reached the scale of geological events.

The fact that information of various kinds was buried (conserved) during the geological history allows us to read the chronicle of evolution of Earth and its biosphere. The buried information (at least a part of it) was accumulated and complicated in the course of evolution (in accordance with the laws of self-development).

Isotope analysis is one of the tools used to determine the age and genesis of geological bodies (and, hence, to study their evolution). Along with paleontology, which gives a relative age, radioisotope dating is used to determine the absolute age (specifically, radiocarbon dating and other kinds of analysis using potassium–argon, uranium–lead, rubidium–strontium, etc.). The observed ratios of stable isotopes of some elements (O, H, C, S,) allow us to draw conclusions about the origin of their compounds, taking into account redistribution of stable isotopes that occurs during different geochemical processes. Separation of stable isotopes is caused not only by their different atomic weights but also by different thermodynamic characteristics (free energy of formation). Therefore, isotope separation can also occur during chemical reactions, being very slow because the ΔG of the reactions differs from unity only slightly. For example, for the isotopic exchange reaction:

$$^{15}NH_3 + {}^{14}NH_4^+ = {}^{14}NH_3 + {}^{15}NH_4^+$$

the equilibrium constant at 25°C equals 1.034.

To date, a huge amount of information has been accumulated concerning separation of stable isotopes of light elements in different geological processes (e.g., see works by Vinogradov).

Let us consider an example of the use of isotopy for figuring out the age of geological events. According to the opinion of another academician sharing the same last name, V. I. Vinogradov (1980), the fact of the existence of Precambrian uranium-bearing conglomerates and Precambrian ferruginous quartzites cannot be taken as evidence that the Precambrian atmosphere was oxygen-free (this is still under discussion). A detailed study of the isotope composition of sulfur in ancient evaporites (sedimentary rocks resulting from evaporation of seawater) allows us to assume that the Earth's oxygen atmosphere dates back to between 3 billion and 3.5 billion years ago:

> From this moment on, sulfur atoms participate in one of the most important phases of the Earth's evolution in the history of geochemistry – the biogenic cycle. The only reason for the substantial fractionation of sulfur isotopes in the natural environment is their redistribution between oxidized and reduced compounds.
>
> (Vinogradov, 1980, p. 176 [in Russian])

Lecture 5

Geochemistry of processes

A prerequisite to any process is the presence of a gradient between values of any parameter (e.g., temperature, pressure, concentration, etc.).

According to Engels, matter has five forms of motion: mechanical, physical, chemical, biological, and social. All of them can be observed in geochemical processes.

Many processes occur concurrently in the Earth's depth. Is there any order in their sequence? Is it possible to develop a system comprising them? What alterations could this system (if there is any) undergo during the Earth's evolution? What changes do the same processes show at different stages of evolution? There are no clear answers to these questions yet.

We know that there are coupled processes, when the completion of one gives rise to the other. Furthermore, there are chains composed of coupled processes. Knowledge of such chains is important for various forecasts, including ecological ones. What was nature's motivation for creating such chains? Do they always have a positive aspect, or are there completely negative ones?

These questions essentially boil down to one thing: because we have stated that Earth is a unified system where all events are interrelated, we would like to know how exactly this interrelation happens and to what extent it is predetermined. Some aspects of this problem have already been solved by the scientists (at least they believe so), but many others are still to be comprehended.

Within Earth, geochemical processes fall into two large groups: (1) endogenous events taking place in the Earth's interior at an elevated temperature and pressure and (2) exogenous processes occurring in the near-surface zone at normal pressure (close to atmospheric) and low temperatures from below zero to slightly above it.

Differentiation of the Earth's matter has resulted (and now proceeds with it) from a variety of geological processes, both endogenous (magmatic, metamorphic, metasomatic, hydrothermal, etc.) and exogenous ones (destruction, sedimentation, evaporation, those of cryogenic or biological type, and others). To a certain extent, the whole set of these processes necessarily undergo progressive evolutionary transformations aimed at reducing entropy, increasing complexity and diversity, and accumulating internal energy. At the same time, geologists recognize the existence of two distinctly manifested and oppositely directed trends: (1) differentiation (separation) of the processes and (2) their mixing and homogenization (dissipation). Both trends are thermodynamically justified. Their combination reveals the principle of unity and struggle of opposites. Newton's third law states that for every action in nature there is an equal and opposite reaction. If the action is not limited, it will "go off scale," and the system will collapse. Therefore, every self-regulating system must have restrictive laws. The role of such a limiting mechanism is played by geochemical barriers that are a special and extremely important factor in regulation of geochemical processes.

A geochemical barrier (Alekseenko and Alekseenko, 2003) is a part of geological space where the element mobility is decreased due to a change in physical, physicochemical, or other factors. Geochemical barriers hamper migration of elements, thereby causing their concentration sometimes to the extent of a basis for mineral deposits, and perform an "ecological" function by purifying natural and technogenic waters and limiting the spread of toxic chemicals. According to Perelman's (1989) classification, the following types of geochemical barriers have been distinguished: acidic, alkaline, oxygen, reducing, evaporative, temperature caused, adsorptive, thermodynamic, complex, and mobile.

The oxygen barrier. A decrease in the mobility of chemical elements at this barrier results from a sharp increase of their oxidative potential. For example, when mineralized underwater comes out on the Earth's surface where its oxidation potential grows sharply due to atmospheric oxygen, ferrous iron undergoes oxidation to its trivalent form and precipitates as a hydroxide. Another example is precipitation of native sulfur in gas fumaroles during hydrogen sulfide oxidation.

The reducing barrier. The following chemicals may act as reducing agents: hydrogen sulfide, both gaseous and dissolved in water; organic matter; and minerals containing elements in a low oxidation state. The presence of sulfide sulfur in the system decreases the mobility of chalcophiles. The so-called black smokers can serve as an example of such a barrier. In some parts of the seabed, vents emit hypogene hydrogen sulfide. Its reaction with metals dissolved in seawater results in the formation of fine-grained sulfides that gradually settle to the bottom. Also, precipitation of poorly soluble sulfides can result from penetration of sulfate solutions containing metals into a region of low oxidation potential. When in their lowest oxidation state, some polyvalent elements become inert and inactive.

The barriers where organic matter acts as a reducing agent (in the absence of hydrogen sulfide) are called gley barriers. They can be seen, for example, in swamps.

The alkaline barrier. This barrier is characterized by an elevated pH of aqueous solutions that result, for example, from their reaction with host rocks. A vivid example of such a barrier is the interaction of acidic solutions with carbonate deposits:

$$CaCO_3 + 2H^+ = Ca^{2+} + H_2O + CO_2.$$

Neutralization of an acidic solution may be accompanied by precipitation of hydroxides of iron or other metals, which is often observed in nature.

The mobile barrier. An alkaline barrier is a convenient example of another type of the geochemical barrier – the mobile one. Imagine that an acidic solution carrying dissolved iron is filtered through a thick rock containing carbonates. On the front line of the solution, the neutralization reaction yields precipitated iron hydroxide and dissolved carbonates. Next, fresh portions of the solution come into contact with yet unreacted carbonate, dissolve the already precipitated iron hydroxide, and the redeposit it further downstream. And this act is repeated many times.

The acidic barrier. This barrier is the opposite of the previous one. In this case, the elements that are mobile in alkaline solutions will precipitate. Realization of such a barrier is possible (e.g., when mixing alkaline and acidic waters). In the mixing zone, both solutions are neutralized, and hence this is a combined barrier. Another example of the acidic barrier is, say, the oxidation zone of a sulfide deposit or the dumps of a mining enterprise that contain sulfide minerals whose oxidation gives sulfuric acid:

$$FeS_2 + 7O_2 + 2H_2O = 2Fe^{2+} + 4SO_4^{2-} + 4H^+$$

The adsorptive barrier. This is a complex and still insufficiently studied barrier that is extremely important for ecology. It works where there is a sufficiently developed and active mineral surface, (i.e., in weathering crusts, soils, bottom sediments, and sedimentary rocks). This barrier is peculiar for its ability to reliably retain sorbed substances for a long time. Apparently, it is mostly due to such barriers that underground freshwater exists. Adsorptive barriers can play both a positive and a negative role in ecology. For example, soils whose good sorption ability results not only from a large specific surface area but also from the presence of organic sorbents (humic acids) absorb all "contaminants" from the solutions that are filtered through them. Then plants take up these contaminants (including toxic ones), thereby introducing them into the biological cycle. Nature is a unified system where all processes are interrelated. Therefore, geochemical barriers are often of a complex type, which is exemplified by adsorptive-alkaline ones that adsorb hydrogen ions on the mineral surface, thereby provoking the increase of solution pH.

Fedoseeva (2003) reported on temperature-induced intensification of chemisorption of a number of elements (Co, Cd, alkaline earth metals) on oxides and a decrease in the intensity of physical sorption of alkali metals. She offered the following series of metals whose hydroxides have different sorptive capacities with respect to cesium:

$$Mg^{2+}, La^{3+} < Be^{2+}, Al^{3+}, Th^{4+} < Fe^{3+} < Sn^{4+} < Ti^{4+} < Nb^{5+}$$

Similarly, for different metals, their cation sorption on the same oxide follows its own pattern: as a rule, it increases with the increasing number of cation charges (single < binary < multiple charges). In Mendeleyev's table, sorption of alkaline earth metals increases from top to bottom.

There are other examples of complex geochemical barriers. We know that oxidation/reduction reactions are often accompanied by acidity/alkalinity changes.

The thermodynamic barrier. Still another complex geochemical barrier is the thermodynamic barrier. It is associated with changes in the main intensive parameters (i.e., temperature and pressure), which can result in alteration of all thermodynamic constants, thereby stimulating or suppressing the action of any other geochemical barrier.

For example, monotonically decreasing temperature of a hydrothermal solution will cause discrete precipitation of minerals, the sequence of which is determined by thermodynamics of the system. This process is complicated by the temperature effect on the dissociation constants of compounds dissolved in water, etc. The thermodynamic barrier is drastically different from any other barrier because it affects not just one parameter of the process but a few of them. A change (for some reason) in the thermodynamic constants immediately puts in joint action the acidic–alkaline, oxidation–reduction, and other geochemical barriers. As a result, all the usual regularities "swim about" and demand a comprehensive analysis.

The action of any geochemical barrier can be described from the viewpoint of thermodynamics; hence, a thermodynamic approach can be the basis for constructing a general theory of geochemical barriers, including complex ones.

The mentioned geochemical barriers function both in endogenous and exogenous conditions.

Endogenous processes

When local temperature is increased due to some reason such as radioactive heat or upward movement of a portion of rocks heading for regions with lower temperature and pressure (because pressure drops faster than temperature, the system may cross the liquidus line), magmatic chambers arise in the Earth's interior. Rocks turn into a melt – the magma. Depending on the system composition, the melting point may vary greatly. Even one component added to the system always makes the melting point lower.

Let us consider this effect using a simple eutectic system (see Fig. 5.1). After the system has been cooled below the liquidus line, either the component denoted by A (if the system composition point is on the left side of the diagram) or that denoted by B (if it is on the right) starts solidifying. Hence, it is not necessarily the most refractory phase that solidifies first. This happens because the process depends not only on crystallization temperature of the "pure" phase, but also on possible formation of a "solution in the melt," where a more refractory component can be retained. In practice, this phenomenon is used when growing crystals of complex compounds.

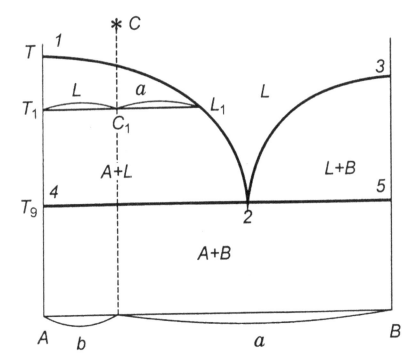

Figure 5.1 Diagram of melting of a simple binary eutectic system.

Here, it is not out of place to make a small "lyrical" digression. In geochemistry, the notion "solution" is usually understood as an aqueous solution. Water is a polar solvent where electrolytic dissociation plays an important role. It seems that geochemistry has practically nothing to do with nonpolar solvents. But this is only on the face of it. Melts (ranging from high-temperature magmatic to low-temperature salt ones) are exactly such nonpolar solvents where the principle "the like dissolves in the like" is realized. From the viewpoint of thermodynamics, there is no fundamental difference between melts and aqueous solutions. In both cases, addition of a new component shifts the phase boundaries and makes the melting/crystallization temperature lower. Therefore, magmatic and hydrothermal processes are conjugated and should be considered together.

After a magmatic chamber has been formed, the melt becomes nonequilibrium relative to its "habitat" and begins to interact with the surrounding solid rocks. This triggers assimilation, which is partial melting of the wall rock. Assimilation is an endothermic process; its consumed energy is compensated by an exothermic process, which is crystallization of a portion of the magma (Le Chatelier's principle). Accordingly, the magma composition changes and the figurative point of the system shifts. Because they are heavier than the melt, the crystallizing minerals settle to the "bottom" of the magmatic chamber. The magma "bites through" the chamber roof and gradually moves upward into the region of lower temperatures. This results in differentiation of crystallization. Depending on the changes in magma composition occurring during wall rock assimilation, crystallization can follow different pathways and yield new rocks different from those melted (Fig. 5.2).

If the phase composition changes along the line 1–2, the result of crystallization will be a mixture of solid phases $A + A_3B$; in the two other cases, the $A_3B:B$ ratio of the mixture will vary.

During pressure reduction, cooling, and magma crystallization, the volatile matter separates from the magma, thereby initiating both metasomatic and proper hydrothermal processes.

Besides, high temperature of the magmatic chamber causes metamorphic transformations of wall rocks (contact metamorphism); that is, the mineralogical composition changes without a change in the total chemical composition.

In the geochemistry of endogenous processes an important role is played by fullerenes, which are specific polyatomic carbon molecules with unique physicochemical properties (Vinokurov *et al.*, 1997). It is believed that fullerenes could take a significant part in the mantle convection and magmatic differentiation, as well as in the formation of diamond deposits and rare-metal carbonatites.

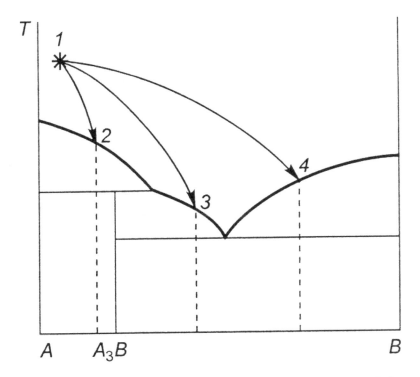

Figure 5.2 Three variants of changes in the phase composition of the system during assimilation and crystallization differentiation.

Having separated from the magmatic chamber, volatile components gradually cool down, first passing the pneumatolytic stage (high temperature – supercritical, 400°C to 600°C) and then the hydrothermal one (from 100°C to 350°C). Hydrothermal processes occurring in cavities of various dimensions and shapes often yield well-cut crystals, which sometimes are quite large.

Hydrothermal solutions provide mobility and migration of many elements due to a high temperature and the presence of various ligands, which ensures that high formation was universally recognized as a dominating factor in transport of heavy metals by aqueous solutions. Subsequent studies were focused on specific types of complexes formed under different conditions by ions and molecules that are present in aqueous solutions. Natural solutions contain a very wide range of potential ligands, including HCO_3^-, CO_3^{2-}, HSO_4^-, SO_4^{2-}, HS^-, S^{2-}, S_n^{2-}, $S_2O_3^{2-}$, Cl^-, F^-, OH^-, various organic compounds, etc. The distribution of various forms of the existence of metals in aqueous solutions depends on the ratio of ligand concentrations, pH, the oxidation potential, and temperature. The problem of complex formation in natural solutions was discussed in the literature by Betekhtin (1953), Bjerrum (1961), Smith (1968), Helgeson (1969), Barnes and Czamanski (1970), Kolonin and Ptitsyn (1974), and many others.

The complex forming–caused increase of the equilibrium concentration of metals in solution is numerically expressed as a complexity value (also called Leden's function), which is the ratio of the total metal concentration to the concentration of its free ions:

$$F = \frac{C_m}{\left[M^{m+}\right]}$$

The complexity value is calculated from the formula

$$F = 1 + \sum \beta_n \, [L^{m-}]^n$$

where β_n is the total formation constant of the n-complex, and L is a ligand.

The complexity value strongly depends on ligand concentration (see Table 5.1), while elevated temperature can both increase this value (in case of weak complexing agents) and decrease it (in case of strong complexing agents).

Thus, thermodynamic calculations allow us to explain high concentrations of ore-forming elements in aqueous solutions, which appear to be much higher than product solubilities of their equilibrium minerals, and to predict the behavior of a particular element in aqueous solution under particular physicochemical conditions.

Table 5.1 The complexity logarithms of ore-forming metals as chlorocomplexes at 25°C (Kolonin and Ptitsyn, 1974)

Metal	Concentration of chlorine ions in solution, g-ion kg^{-1}					
	10	4	1	0.4	0.1	0.01
Cu$^+$	8.17	7.01	5.35	4.36	3.00	0.99
Cu^{2+}	1.57	0.94	0.35	0.16	0.04	–
Ag$^+$	9.54	7.99	5.83	4.69	3.34	1.59
Au$^+$	11.0	10.2	9.0	8.2	7.0	5.0
Au^{3+}	30.0	28.4	26.0	24.4	22.0	18.0
Zn^{2+}	3.89	2.36	0.55	0.14	0.02	–
Hg^{2+}	20.3	18.8	16.4	14.9	12.07	9.63
Pb^{2+}	6.10	4.59	2.61	1.71	0.88	0.19
Fe^{3+}	4.37	3.39	2.10	1.36	0.52	0.06

As the solutions move up toward the Earth's surface, temperature decreases, the instability constants of complex constituents vary, and the cations "change their partners."

During the hydrothermal process, the sequence of mineral crystallization depends on the medium composition and thermodynamic constants of the complex members (i.e., ions and molecules). The idea that this sequence is determined solely by product solubilities of minerals is erroneous, even though some geochemists believe it to underlie mineral zoning typically observed in hydrothermal processes. Such a typification is explained by the fact that nature repeats the same geochemical conditions, but the reasons for mineral zoning are more intricate and complex.

Hydrothermal solutions can be formed not only as a result of separation of volatile components from the magmatic chamber, but also from the so-called meteoric hydrotherms resulting from surface water immersion in the deep crust (e.g., during marine sediment burial). At a certain depth where temperature and pressure reach the hydrothermal level, the solutions become chemically active, and a metasomatic or actually hydrothermal process is initiated.

The metasomatic process is a variant of the hydrothermal process. One of its peculiar features is that on its way through a porous rock the hot solution interacts with this rock, dissolving some elements and making deposits of others. Specifically, during this process the porous rock plays the role of a membrane where, due to the membrane effect, the elements are separated. This phenomenon has been experimentally demonstrated by Korzhinsky's progeny.

Another peculiar feature of the metasomatic process consists in surface events whose role in the finely dispersed medium increases sharply. The interaction of solutions with small-size particles is drastically different from that of the "water–rock" type in macrosystems. The surface events represent a particular section of thermodynamics; they have been described in detail by Rusanov (1967).

Yet another feature of the metasomatic process was predicted by Korzhinsky and later shown by his progeny. Namely, it is the acidic wave that accompanies fluid filtration through porous rock. It arises because protons due to their small size and higher mobility outstrip the larger hydroxyl ions. As a result, the front portions of the solution become acidified in contrast to the rear ones that are alkalized. Hence, a mobile geochemical barrier arises that works as follows: first, the acidic solution dissolves acid-unstable minerals, and then, with increased pH, minerals that are poorly soluble in a neutral medium undergo crystallization.

Rapid mass crystallization in a hydrothermal solution may be driven by the throttling effect; that is, when a crack opens, its volume instantly increases, thus triggering a rapid temperature decrease. In this case, the minerals do not comply with the prescribed order of crystallization, and the result is a chaotic mixture of typically small crystals.

Lecture 6
Geochemistry of processes (continuation)

Exogenous processes

The main types of exogenous processes are presented in the classification developed by Fersman (1959):

1 Hypergenesis as such, which is transformation of crystalline rocks and ores. Formation of weathering crusts, oxidation zones, etc.
2 Pedogenesis, which is soil formation on dry land.
3 Syngenesis, which is sediment formation in waters.
4 Diagenesis, which is transformation of sediments resulting in their compaction.
5 Catagenesis, which is a set of geochemical and mineralogical processes occurring in sedimentary rocks at elevated temperatures and involving water, CO_2 and other agents.
6 Halogenesis, which is precipitation of salts from aqueous solutions.
7 Hydrogenesis, which is a set of processes caused by penetration of natural waters into the lithosphere (crust, etc.).
8 Mechanogenesis, which is the processes of mechanical transport and deposition of matter.
9 Biogenesis, which is a set of biogeochemical processes.
10 Technogenesis, which is a set of geochemical processes caused by various human activities.

In the days of Fersman, the permafrost was considered to be a zone of chemical rest. Today, everyone knows that this is not so. Therefore, a necessary addition to this classification is cryogenesis, a set of geochemical processes occurring at negative temperatures (Celsius). Also, atmogenesis includes processes occurring both directly in the atmosphere and on its boundaries with the hydro- and lithosphere.

The exogenous processes are discussed in detail in the geochemical literature (Perelman, 1989, Alekseenko, 2000 and others). Here, we will focus only on processes most significant for global geochemistry, namely, sedimentation and element cycling within Earth.

Annually, the World Ocean receives more than 20 trillion tons of sediments; more than 80% of them are terrigenous (solid, arriving with water and wind). Mostly, they come through deltas of the largest rivers, where deposit thickness may exceed 10 km. As to fine material, it is actively "supplied" by volcanoes and wind erosion. For example, between 60 million and 200 million tons of dust are yearly carried away from the Sahara area. After compaction and cementation, the sediments undergo lithification and finally turn into a dense low-porous rock.

Some modern geologists believe that all sedimentary rocks are of biogenic origin, and that the granitic magma comprises "former" biospheres. This statement is not indisputable because there are purely inorganic sedimentary materials (e.g., terrigenous volcanic material, etc.). On the other hand, it is the biosphere that "controls" the cycle of elements (including water circulation) in the Earth's interior, where they return in no other way but through sedimentation. The biota-accumulated solar energy also uses this way. Additionally, the sedimentation cycle removes some carbon (in the form of kerogens) from circulation, thus maintaining a certain oxygen concentration in the atmosphere, preserving the ozone layer, and reducing the greenhouse effect. So, the sedimentary processes should be recognized as a most important factor of the geochemical "life" of the planet.

The global geochemical cycle

Against the background of the general irreversible progressive development of the Earth's crust, different geochemical epochs showed common features due to repetition of the following processes: underwater volcanism, marine sedimentation, metamorphism, intrusive magmatism, land volcanism, erosion, weathering crust formation, continental sedimentation.

First proposed by Vernadsky, the concept of geochemical cycles now dominates in geochemistry and considers tectonic processes, magmatism, sedimentation, and life evolution as constituents of a unified development process. The geochemical cycle should be understood as a form of combined rectilinear and rotating motion which can be illustrated by a cycloid or a helix. In general, the duration of major tectonic magmatic cycles correlates with the galactic year (time required for the Sun to orbit once around the center of the Milky Way Galaxy), which is equal to about 180 million to 220 million years.

It is established that each relatively short cycle has the same basic features of evolution as those of a larger cycle. Rundqvist (1965) called this dependence a geogenetic law and demonstrated its force in some geological processes.

According to the present concepts (Perelman, 1989), at its initial and final stages each major geochemical cycle is characterized by sea regressions, mountain building, abundantly arising arid landscapes, reduction of biomass, a decrease in the amount of organic carbon in sediments, and decreasing carbonate accumulation in the seas. In these periods the role of living matter lessens. It is believed that the periods of mountain building gave rise to new species, genera, and families; that is, an "explosive speciation" took place. Conversely, the middle stages of the cycles are characterized by significant sea transgressions, peneplanation, a mild and humid climate, growing biomass, accumulation of organic carbon in sediments, active volcanism, and carbon dioxide inputs to the atmosphere. Figure 6.1 shows the global geochemical cycles over the last 600 million years of the Earth's history and concurrent changes in concentration of atmospheric carbon dioxide (the data obtained from paleoreconstruction). As seen, the carbon dioxide concentration in the atmosphere increases in the middle of the cycles.

Activation and inactivation of processes occurring in the Earth's interior correlate with the cyclicity of gravitational energy transferred to Earth from cosmic

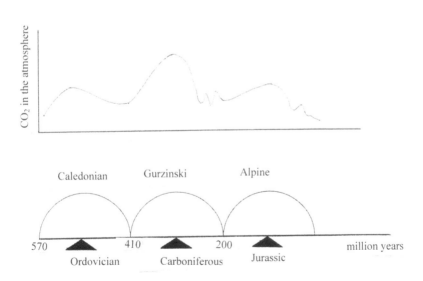

Figure 6.1 Global geochemical cycles of the last 600 million years and the changing concentration of atmospheric CO_2 (upper curve).

bodies (the galactic year). The chain of dependencies in the global geochemical cycle can be easily traced: (1) When the Earth's energy is spent on active volcanism – the concentration of atmospheric CO_2 increases – plant nutrition improves – life becomes more comfortable (more food for all species) and the need for new species decreases – biomass grows – more organic carbon is deposited. (2) When volcanism weakens (Earth's energy is spent on mountain building) – the concentration of atmospheric CO_2 decreases – biota has nutrition problems – biomass is reduced – new species of higher resistance are needed – speciation is intensified. This is how geological processes affect the biosphere.

But the biosphere, in turn, affects geochemistry and energy of the Earth's depths. In the course of matter convection in the mantle, the composition of the descending branch is determined by processes occurring on the Earth's surface. Consequently, the cyclicity of biosphere processes determines cyclic changes in the composition of sediments sinking into the Earth's interior, as well as the cyclic release in the mantle of energy stored in organic matter, which, in turn, ensures a certain cyclicity of magmatic and other endogenous processes.

Thus, periods of active volcanism are favorable for the formation of sediments rich in organic matter which can later turn into deposits of coal, oil, gas, and other raw materials. Consequently, it is very likely that the formation of hydrocarbon fields was of cyclic nature. Because all the interior processes are interrelated, it seems possible, although with a lower probability, that deposition of other minerals, ores in particular, is a cyclic process as well. It should be remembered that organic matter contains not only biogenic elements, but also many others that at a certain stage of the global geochemical cycle can be involved in endogenous processes (e.g., hydrothermal) and become a part of an ore deposit.

Let me remind you the above quoted opinion of some scientists that "the granitic magma consists of former biospheres." For reference only, the volume of water passed through living matter of the planet during the last 500 million years is 50 times larger than the entire hydrosphere of the Earth, and the total mass of matter passed through living organisms during the entire geological history is comparable to the mass of Earth. So, living matter seems to "have digested" the whole Earth, although it cannot be said that every atom of terrestrial matter had been involved in biological cycles.

The geochemical cycle comprises processes alternatively accompanied by increasing or decreasing entropy. Those with increasing entropy dominate deep in the Earth (foci of magmatism and metamorphism), whereas those with decreasing entropy are typical for near-surface areas, especially the biosphere. Hence, endogenous processes "prefer" equilibrium thermodynamics, while nonequilibrium thermodynamics is favorable for exogenous events.

The doctrine of geochemical cycles serves as a core tool in describing the evolution of geochemical systems. Geochemical cycles of different taxonomic ranks (including the water cycle and the biological cycle of atoms) provide the circulation of Earth's elements and preserve the atmosphere and hydrosphere compositions within certain historically settled limits. Violation of these cycles can cause unpredictable environmental situations.

Therefore, as a geochemist-ecologist, I consider it expedient to include in the criminal law an article on "violation of the laws of nature resulting in irreversible changes in ecosystems."

From the viewpoint of geochemistry and ecology, endogenous and exogenous cycles of carbon are of great importance. The carbon content of an exogenous cycle is three to five times lower than that of an endogenous cycle, but the former is much shorter, so their contributions are comparable. In an exogenous carbon cycle, one of the key processes is formation of a stable organic matter from the dead biomass.

In the carbon cycle, a significant role is played by methane, which is the second most important greenhouse gas, with an effect 20 times higher (per molecule) than that of CO_2. The natural flow of methane is estimated at about 2 million tons per year on oceanic shelves and about 8 million tons per year in abyssal zones.

In nature, methane is most intensively supplied by swamp; its flow on the bog surface is extremely uneven both in time and space. Almost 60% of the world's methane is believed to come from anthropogenic sources, such as rice fields, cattle digestion, biomass burning, gas production, etc.

The geographic (exogenous) cycle of water includes its evaporation from water bodies (mostly oceans and seas), coming back to the Earth's surface by precipitation, and movement in the form of rivers and glaciers. The average residence time of water molecules in the atmosphere is estimated at 11 days.

The large (exogenous/endogenous) water cycle includes the following stages (Shvartsev, 1998). Atmospheric water with a certain amount of dissolved carbon dioxide precipitates. On the Earth's surface, this water interacts with clays that are weathering products of crystalline rocks. Due to their high adsorption properties, clays bind a significant number of H^+, thereby yielding clay acids and promoting dissociation of water:

$$H_2O + clay \rightarrow H\text{-}clay + OH^-$$

The interaction of hydroxyl-ion with carbon dioxide gives bicarbonate-ion

$$CO_2 + OH^- = HCO_3^-$$

that reacts with metals dissolved in water (Ca, Mg, Ba, Fe, etc.) and forms carbonate deposits in the sedimentation zone:

$$Ca^{2+} + HCO_3^- = CaCO_3 + H^+$$

The released H^+ can either bind to clays or participate in other reactions. So, in the hypergenesis zone water can be completely decomposed into hydrogen and oxygen. The formed clay-carbonate sediments slowly sink into the Earth's depth where high temperatures and pressures make them decompose and release water vapor and CO_2. Thus, it is in this deep zone where water is synthesized. The further pathway of the released gases lies through either metamorphogenic waters or the magmatic process with subsequent hydrotherms and volcanism that carry H_2O and CO_2 onto the Earth's surface.

As seen, the water cycle is related to one of the branches of the carbon cycle. Another branch of the hypogene carbon cycle is associated with the biological cycle of atoms. Having plunged deep into the Earth, metabolic products of various organisms may undergo either complete or incomplete decomposition. Their complete decomposition gives inorganic components (CO_2, H_2O, Ca, Na, Mg, K, S, etc.) that in the course of geological processes are carried onto the Earth's surface where they are consumed by photo- and chemosynthesis to yield living cells and free oxygen.

Incomplete decomposition of metabolic products results in formation of organic soil components, silts, and sedimentation rocks, including coal, peat, oil, and gas. The formed organic matter can be kept away from the cycle of atoms for a long time, but when moved again into a zone of high temperature and pressure, it undergoes mineralization and gets involved in the circulation.

It is obvious that water is a member of the biological cycle of atoms. Together, all this is evidence for the diversity and interrelationship of matter's cycles in nature.

It can be assumed that similar to how largely the macro behavior of elements in physical, chemical, biological, and technological processes is determined by their atomic and nuclear properties, the roots of macroscopic oscillation processes lie in the wave nature of matter. Interference of coupled oscillatory processes in the course of evolution allows them to form cycles of different lengths, thereby stimulating or suppressing the progressive development of geosystems at different stages of evolution. Thus, geochemical cycles can be of both outer (cosmic) and inner origin.

The carbon cycle holds the central position, whereas cycles (circulations) of other elements are coupled with it on terms of clear quantitative relationships. For the purpose of oxygen accumulation (and maintaining it at a proper level) in the atmosphere, it is necessary to remove from it some carbon, which is performed through formation of carbonates and kerogens (coal, oil, gas, etc.). This is possible only under conditions of incomplete oxidation of organic metabolic products. Thus, the formation of hydrocarbon deposits is one of the goals of the biosphere. Here again humans oppose the laws of nature.

Geochemical cycles combine cyclic movements of chemical elements within Earth which follow not a circle, but a helix (or cycloid). Each next turn, although preserving the basic features of the previous one, can have its own peculiarities, such as dead-end branches where a part of matter is removed from the cycle. This is how nature lays in a stock of substance and energy for the future.

Less than 1% of the total carbon amount is involved in the immediate circulation between the atmosphere, soil, oceans, and living organisms.

The bulk of Earth's carbon is stored in the form of carbonate rocks and dead organics – fossil fuel.

The interaction of water, carbon, and oxygen cycles occurs in conditions of a complex balance between its participants.

According to James Lovelock and Lynn Margulis, the main regulators of biospheric processes are anaerobic bacteria that do not allow the level of free oxygen to rise above the permissible value (25%). Exceeding this level will lead to spontaneous fires.

So, we see that nature has worked out a system of mechanisms united in a legislative base – a catechism regulating the "behavior" of "citizens" of the planet Earth. Of course, this catechism cannot cover all situations, so there are calamities which Earth has to fight. One of these is fire that can hardly be thought useful.

Forest fires are recognized as one of the most important factors in geochemical transformation of landscapes. Long-term studies of fire sites in various regions of Siberia (Scherbov *et al.*, 2015) distinguished two groups of elements with different patterns of behavior in fires: (1) active air migrants (Hg, Cd, Pb, Zn, Mn, As, Sb, Se, U, and artificial radionuclides), and (2) passive elements accumulating on fire sites (Cr, Ni, Co, V, Mg, Si, Fe, Th, Ca, K).

General crown fires result in redistribution of elements over geochemical landscapes, emission of dust and aerosols in the atmosphere comparable in scale to volcanoes, significant transformation of landscapes, and destruction of forests (the core of landscapes). Fires naturally generate new geochemical processes.

Geochemical activities of mankind were termed "technogenesis" by Fersman. Expending a large amount of energy, people accelerate natural processes and create new ones that are not characteristic to unexplored wilderness. The result is an increased differentiation of matter, negentropic tendencies in geochemical migration, and elevated temperature of technogenic systems in comparison with natural ones (temperature of cities and industrial centers, and temperature pollution). All this adds to global warming.

Thus, the Earth's geochemical processes are not individual but interrelated with one another through direct and reverse connections, both positive and negative. The study of their chains is important both as a geochemistry goal and the basis of ecological forecasts. Also, it won't do any harm to finally address the system of geochemical processes and the laws that govern it.

Lecture 7
The biosphere

The biosphere is the worldwide sum of all living beings, the zone of their life on Earth and their relationships, including their interaction with the nonorganic elements of the lithosphere, hydrosphere, and atmosphere (Ptitsyn, 2013, Alekseenko, Alekseenko, 2002). This is what we take for the exclusive boundaries of the biosphere. Nature exploration and discovery of novel species contribute to the biosphere expansion.

The possibility of life on Earth is determined by the Earth's position in the solar system (distance from the Sun) that allows the presence of liquid water, which is the cradle of life and its natural habitat. Incidentally, the same criterion – the presence of liquid water – is considered when discussing the possibility of life on Mars.

As the most important geochemical factor and the major component of the biosphere, water deserves special attention. It is involved in the majority of events occurring in the Earth's crust.

At present, oceans and seas occupy 71% of the Earth's surface. Water is a necessary constituent of human life and the biosphere as a whole. On average, the water component of animals and plants amounts to 50% (for humans, 65%). Water is needed in industry, transport, and energy production. The daily consumption of water is estimated at about 7 billion tons.

Water properties support the biosphere's evolution in all respects. Its boiling and thawing points set the temperature boundaries of life for the majority of organisms; its heat capacity, fusion heat and heat of vaporization regulate the climate; ice that floats in its own melt protects reservoirs from freezing; oceans play an essential role in maintaining the global gas regime; due to its high dielectric constant, water is an inert solvent fit for supplying nutrients and removing wastes from organisms without entering into a reaction with their tissues. The lack of any of these water functions is detrimental to life.

Water is an abnormal compound. According to Mendeleyev's periodic law, water should be an analog of hydrogen sulfide, hydrogen selenide, and hydrogen telluride. Had it been so, in the conditions of the Earth's surface water would have been a toxic gas with an unpleasant odor, with a freezing point of about $-100°C$ and a boiling point of $-50°C$. Abnormal properties of water (high values of heat capacity, latent fusion and evaporation heats, dielectric constant) result from structural aggregates (tetrahedra, chains, rings) formed by H-bonding. The structure of water, and hence its properties, can change under the influence of external factors. For example, simultaneous exposure to high temperature (up to 400°C) and high pressure (up to 100 MPa) or slow ice thawing without stirring gives activated water, which is characterized by a reduced pH, elevated electrical conductivity, and increased dissolving power for many minerals. Also, meltwater shows changed physical properties (e.g., the value of the dielectric constant). An electric field weakens the structural bonds of water. A magnetic field enhances the dissolving abilities of water; therefore, instead of toxic reagents, activated water is used, for example, to remove scale from boilers of thermal stations.

The sensitivity of water properties to environmental effects was also revealed in studies of natural geophysical fields. Some researchers believe that water properties, including its structure, have varied in the course of the Earth's evolution. The rate of some chemical processes has been shown to depend on oscillations of the intensity of the Earth's magnetic field. Water possesses a structural memory, that is, the ability for a long time (up to a few months) to retain the properties that resulted from its exposure to physical fields.

The thermal motion of molecules in liquids may be characterized as ambivalent. On the one hand, it is oscillation about a point of equilibrium, and on the other hand, it is a jumplike transition from one equilibrium state to another. Such a transition is called translational; it loosens the water structure. The ions of inorganic compounds dissolved in water affect water structure in different ways. Some ions (structors) hamper the translational motion of water molecules, thereby strengthening their structure. These include small single-charged ions and multiply charged ones, that is, ions with a relatively high energy constant (EC): Mg^{2+}, Ca^{2+}, Ba^{2+}, Na^+, Li^+, SO_4^{2-}, etc. The other group (destructors) includes large single-charged ions with a low EC that stimulate the translational motion of water molecules, thereby loosening their structure.

The effect of dissolved compounds on the water structure is comparable with that of temperature and pressure. In 1944 John Desmond Bernal and Sir Ralph Howard Fowler proposed the concept of structural temperature of a solution, and later Kopeliovich introduced the notion of structural pressure. Structural temperature is defined as temperature at which pure water has the same structure as a solution of a certain composition and concentration. Structural pressure is defined similarly.

According to Vernadsky, living matter is the totality of organisms on the planet (or on a part of it) expressed in terms of mass and energy. For example, the geochemical activity of living matter is a common feature of soil, sedimentary rocks, the oceans, and groundwaters. Consequently, living matter is the central basic component of the biosphere.

Life on Earth naturally originated the biosphere – a powerful, evolving, self-developing geosphere that performs the critical function of formation and maintenance of a system of element circulation in the planet. It was Vernadsky who elucidated the role of living matter and the geochemical functions of the biosphere. Today his concepts have developed into biogeochemistry, a significant branch of science. Circulation of matter is necessarily required for normal functioning of the biosphere, that is, for supplying food to the habitat and removing metabolic products from there. The driving force of this process is a much higher rate of metabolic processes in bioinert systems as compared to inorganic systems. In turn, this results from some peculiar properties of living matter that are unobserved in inanimate matter.

The "pressure of life," its striving for extended reproduction, underlies a geochemical superiority of life on the planet. The living matter mightiness (according to Vernadsky) results from its instinct of self-preservation, which is not and cannot be inherent to inanimate material.

The rate of biological reproduction is fantastic. Vernadsky wrote (1954, pp. 184–185 [in Russian]) that "for cholera bacteria, the rate of transfer of the geochemical life energy is approximately 33,000 cm/s." Had there been no restrictions, small ordinary infusoria (*Paramécium caudátum*) could have given in five years a mass of protoplasm as large as the 104-fold Earth's volume. But the restrictions exist. Nature has developed laws and mechanisms responsible for a "rational" dynamic equilibrium. This led Vernadsky to the conclusion that the amount of living matter on Earth remained comparatively constant over a long geological time. Traces of an active organic life, probably in its earliest forms, are clearly seen in the isotope composition of sulfur from Precambrian metamorphic rocks (Vinogradov, 1980). Geological studies demonstrate a comparative invariability of the chemical composition of the Earth's crust throughout the entire geological time (Vinogradov, 1980).

Worldwide, a molecular genetic analysis of microorganisms is used to study the history of biosphere evolution. Let us refer, for example, to the paper by Battistuzzi, Feijao, and Hedges (2004), titled "A genomic time-scale of prokaryote evolution: insights into the origin of methanogenesis, phototrophy, and the colonization of land." The events were dated using the local clock method (Kumar, 2005). The authors reconstructed the evolution of prokaryotes (Table 7.1) using the sequences of 32 proteins common to 72 prokaryotic species from all the major groups.

Table 7.1 The history of biosphere evolution

Event	Dating (billion years ago)
Origin of life	< 4.1
Origin of methanogenesis	3.8–4.1
(CO_2 + H_2 = CH_4 + H_2O + energy)	
Divergence of the major evolutionary lineages of Archaea	3.1–4.1
Origin of anoxic photosynthesis	< 3.2
Origin of anaerobic methanotrophy	> 3.1
Colonization of land	2.8–3.1
Origin of cyanobacteria	2.6
Origin of aerobic methanotrophy	2.5–2.8
Divergence of the major evolutionary lineages of Bacteria	2.5–3.5

Evolution of the biosphere

Concerning the evolution of the biosphere, especially its early stages, different opinions are expressed, although the main scheme is recognized by most researchers.

Some geologists believe that the oxygen atmosphere originated due to the Earth's degassing upon its cooling, whereas Zavarzin and other microbiologists think that at the dawn of Earth's existence the major role in oxygen production was played by cyanobacteria. Zavarzin (2003b) considered bacteria to be the basis of the biosphere with a superstructure formed by all other organisms. He wrote that, cyanobacteria do not need evolution, they are self-sufficient. They do not have trophic chains, thy do not eat each other. But their geochemical role is enormous. According to Zavarzin, the appearance of the superstructure did not cause any increase in the geometric and physicochemical parameters of the biosphere. The limits of tolerance of higher organisms are much narrower than those of bacteria. Microorganisms extend the biosphere to places where other organisms cannot live. Many things are known about their geochemical activity, but, of course, not all of them.

The "peaceful" life of the evolving biosphere was disturbed by a number of catastrophic collisions of the planet with cosmic bodies (comets, large meteorites). Today, more than 230 large impact craters whose diameters amount up to 200 km are known on Earth. During the last 250 million years, at least nine cases of mass animal extinction have been identified, including the famous Cretaceous extinction (65 million years ago) when about 90% of all species died out.

Space occurrences (Zadonina *et al.*, 2007) caused a chain of important events on Earth: climate change accompanied by a rearrangement of biodiversity; activation of volcanism leading to changes in the atmosphere; strong gamma-ray bursts causing ozone splitting and fall of temperature; and the appearance of special impact rocks, impactites, with fritted grains and a high content of iridium – an element of cosmic origin.

Climate (from the Greek *klima*, meaning "inclination") is to a certain extent a geochemical factor as well, because it affects the mobility of chemical elements in the biosphere. In the past, changes in global climate were caused by cosmic phenomena and led to sharp, sometimes even catastrophic, changes in the biosphere. Currently, the human impact on global climate (mediated by the greenhouse effect) is being widely discussed. In this matter, science is faced with conjuncture and politics. No doubt, climate is affected by human activities, but the degree of this impact is apparently overestimated. Let us recall, for example, the antifreon campaign with accusations of destroying the ozone layer. In fact, the thinnest ozone layer is observed over Antarctica and Eastern Siberia, and these regions cannot be said to use too many refrigerators.

The most important components of the biosphere are the ocean and soils (Ivanov, 1996). In its capacity of a system, the ocean acts as a buffer in various geochemical cycles. It regulates the amount of oxygen and carbon dioxide in the atmosphere and plays an important role in the carbon cycle. Its salt balance depends on the following factors: incoming precipitation from the continents, underwater volcanic activity, solubility of salts in water, substance exchange with the atmosphere, sedimentation, and vital functions. On the ocean floor, mineral deposits are formed both at shallow (shelf) and great depths. Shelf deposits are fine sediments rich in gold, uranium, platinoids, and other elements. Also, shelf areas are rich in hydrocarbons. At great depths, very rich deposits of ores have been discovered, for example, those having a very high cobalt content unobserved anywhere on land. The ocean also is of vital importance for climate regulation.

Soil is the life-giving basis of terrestrial biota, and therefore it is much better studied than the ocean. However, researchers from the Institute of Soil Science still find something to do. Dokuchaev, the father of soil science and Vernadsky's teacher, defined soil as the upper layer of rocks on the immediate surface of the Earth that have been naturally altered by water, air, and living organisms. The main factors of soil formation are the geological substrate composition, bacteria, plants and animals, climate, relief, soil water and gases, time, and human economic activity. Soil is a fine-grained substance comprising clay particles and humic acids that have high sorption ability. Therefore, soils are rich in deposited uranium, mercury, and other toxic elements.

This course of lectures is too short to talk about the biosphere in sufficient detail, so we will dwell only on one more of its specific features.

The biosphere has some special "hidden" areas. Let's name their totality the crypto biosphere (from the Greek *kryptos*, meaning "secretive").

The crypto biosphere comprises the "outskirts" of the major biosphere that are inhabited solely by microorganisms. These include nonfreezing solutions of the permafrost zone; cold groundwater isolated from the major biosphere; and millimeter-size pore water. Specific properties of water in these kinds of extreme systems are determined by a specific composition and particular functions of the microbial community living there. Soda lakes may also be conventionally grouped with the crypto biosphere; according to Zavarzin, they are relics of the protobiosphere that can be regarded as working models of the early biosphere. However, soda lakes are inhabited not only by cyanobacteria but also, for example, by algae.

The permafrost zone covers as much as about 60% of the territory of modern Russia. The inside temperature does not exceed $-10°C$ (except for the surface layer, termed the active layer). Due to a lower freezing point of water in solutions and film moisture (resulting from a reduced chemical potential of H_2O), the liquid water phase is virtually always present in the permafrost. This allows active physicochemical and biological (microbial) life of the cryosphere, because psychrophilic microorganisms (psychrophiles, from the Greek *psychros*, meaning "cold," and *philia*, meaning "love," "addiction") are capable of growth and reproduction in cold temperatures ranging from $0°C$ to $-10°C$. Furthermore, high mineralization of the solution cannot inhibit them from living and developing there.

The life of microorganisms in film water has been almost not at all studied. Meanwhile, the properties of adsorbed water differ much from those of ordinary normal moisture. The thickness of aqueous films adsorbed on mineral surfaces is too small even for microorganisms (up to 20 nm). However, the formation of microbiophilic films (supramolecular ensembles) comprising bound water cannot be ruled out, thus granting the existence of a specific microbial community in millimeter-size pores. This issue has not been addressed at all. It requires joint efforts of geochemists, physicists, chemists, and microbiologists.

Many soda and salt lakes are highly productive extreme ecosystems. A significant seasonal contribution to the functioning of the energy-consuming sulfur cycle of these ecosystems can be made by microbial communities of the photic and wet sandy coastal zones that convert organic matter and minerals. The involvement of microorganisms in the sulfur cycle at the dawn of the biosphere once again emphasizes their crucial role in geochemical processes.

The specific "population" of the crypto biosphere is also of interest from the point of view of the mechanisms of organism adaptation to extreme environmental conditions, and specifically, for exploring the feasibility of life on other cosmic bodies (e.g., Mars).

One more important feature inherent in microorganisms should be mentioned here. It is their ability to cleave a great variety of chemical compounds. Some scientists called it "pantophagy," meaning that in principle there can exist a microorganism capable of oxidizing in certain conditions any substance theoretically prone to oxidation.

These are the general biological grounds for considering the ability of bacteria and microscopic fungi to cause damage to materials as one of the fundamental problems of residence in long-term space stations.

An important stage in the biosphere evolution is the origin of man (Golubev and Shapovalova, 1995), the only biological species that by its own will can determine the path of its development. This "arbitrariness" of man has led to a number of negative consequences. According to Vernadsky, by the beginning of the 20th century human impact has reached the scale of geological processes. By the middle of this century the negative human impact upon nature has provoked the origin of ecology in the broadest sense of the word and submission of environmental problems to large-scale world conferences (Rio de Janeiro, 1992; Johannesburg, 2002). Accordingly, "ecological geochemistry" has emerged as an increment in science. At last people have begun to think seriously about the future of mankind. The first global model of the world development was presented by Meadows and colleagues in the book *Limits to Growth* (1972). Although this model is not directly related to geochemistry, it is immediately related to the Earth's evolution.

Lecture 8

Environmental chemistry

Chemical pollution

The modern environment is a result of coactivity of nature and humans (Bgatov, 1993; Ecogeochemistry . . ., 1996; Hentov, 2005; Solov'ev and Solov'eva, 2013; Vakhromeev, 1995). Comfort and safety of people and other organisms in a particular residential area are legislatively established by the empirically found maximum permissible concentration (MPC) of polluting substances in water and soil and the maximum permissible emission (MPE) of polluting substances in the air. These characteristics vary from country to country and are subject to recurrent modification. MPC shows dangerous concentrations of each particular element, but the environment often contains a few toxic elements at once. Then the following formula is used to estimate their total toxicity:

$$K_o = \Sigma \frac{Ci}{MPCi}$$

where C_i is concentration of the ith element.

Unity represents the upper limit to this parameter. It should be noted that neither this restriction nor MPC take into account the possible effect, stimulating or inhibitory, of one substance on toxicity of another. So, the MPC system as a criterion for the quality of natural media is most imperfect, although it has been legislatively approved. This system is convenient for decision-makers, because it requires no analysis or interpretation. However, no alternative system has been proposed yet.

Because the term pollution is used in the context of environmental problems, it would be logical to take the limits of ecosystem tolerance as criteria for pollution. In other words, the technogenic chemical impact on the environment can only be called pollution when it leads to irreversible changes in ecosystems. Otherwise, the term technogenic pressure would be more correct as it describes intensity of the technogenic chemical impact per unit area of the Earth's surface.

The technogenic pollution affects all four substances forming the residential and life-supporting systems, that is, the atmosphere, hydrosphere, biosphere, and lithosphere. From the ecological point of view, the most dangerous are chemical technologies, especially those involving electrolysis, pyrometallurgy, and burning of solid fuel; this group also includes some agricultural technologies that use toxic organic substances.

The United Nations–approved list of major pollutants of the atmosphere includes CO, H_2S, SO_2, dust, NO, NO_2, hydrocarbons, Hg, Pb, Cd, chlorinated organic compounds (DDT, etc.), oil, nitrates, nitrites, ammonia, and microbial contaminants. These pollutants come to the atmosphere from a wide variety of sources, such as metallurgical, chemical, and nuclear industry; thermal and cement plants; transport; different purpose explosions; fires; losses in oil and gas production; plus degassing of contaminated lithosphere and fumes of contaminated waters.

In the hydrosphere, the major polluting agents are chlorides, sulfates, nitrates, heavy metals (especially Cu, Pb, Cd, Hg, Cr, Ni), hydrogen sulfide, oil, phenols, pesticides, and other agricultural wastes. They come from industrial enterprises that use water in their technological cycles or discard/store solid wastes with toxic elements that can be leached away by the atmospheric and surface water.

Solid substances, such as the lithosphere and soil, are contaminated mainly through the atmosphere and the hydrosphere.

As an alternative to the normative approach to environment quality assessment based on MPC systems and other related indicators, the concept of environmental risk for human health was proposed. It is based on the premise that the constant presence of potentially harmful substances in the environment creates a certain level of real risk. Its supporters claim that there is a limit of nature-conservative measures that, when achieved, ruins the basic production economy. However, this concept has not received wide practical application because in each case it requires a special long-term analysis.

Lastly, the third alternative – the concept of environmental regulation – implies the development of regulations for anthropogenic impact on the environment the observance of which would guarantee normal functioning of ecosystems. This approach is most logical and science based because it follows the laws of nature. However, it is not free of anthropocentric features either, because the norm is understood as a qualitative transitions-limited domain of existence of ecosystems that satisfies the existing ideas of a comfortable residential area for people without taking into account the interests of other inhabitants of the biosphere. Besides, this concept cannot be applied widely because it requires great manpower resources and high expenditures.

As mentioned above, the atmosphere acts as a source of pollution for other components of the environment, that is, the hydrosphere, soil, and the lithosphere. As an example, let us consider acid rain. In the atmosphere, the major acid components are sulfur and nitrogen oxides that are emitted in huge amounts by industrial (primarily metallurgical) enterprises. Their interactions with moisture and oxygen of the atmosphere give the following acids:

$$2NO + H_2O + 0,5\ O_2 = 2HNO_2$$

$$2NO_2 + H_2O + 0,5O_2 = 2HNO_3$$

$$SO_2 + H_2O = H_2SO_3$$

$$2SO_2 + O_2 = 2SO_3$$

$$SO_3 + H_2O = H_2SO_4$$

This acidic mixture is carried by air flows and falls to the ground with atmospheric precipitation, which has various negative consequences, such as soil deoxidation, which worsens its fertility, and extensive damage caused to terrestrial vegetation, as well as to buildings and monuments in cities. Acid rain interactions with dumps of mining and metallurgical enterprises result in leaching away heavy metals and other toxic components, which then penetrate into surface water and groundwater. When getting onto the ocean surface, acid rain causes the death of sea inhabitants.

Every year the atmosphere delivers to the near-surface geosystems millions of tons of technogenic dust with adsorbed highly hazardous substances, such as mercury, radioactive nuclides, etc.

The most active suppliers of radioactive elements to the atmosphere are solid fuel thermal stations because coal often contains tangible amounts of uranium. Once I happened to make the environmental impact assessment at a thermal station that used coal from a uranium-bearing province. A simple calculation showed that this station annually emitted about 9 tons of uranium from its chimneys despite the fact that the filters installed there catch 95% to 96% of dust and gases.

Atmospheric pollutants are a frontier-neglecting disaster of the planetary scale.

The major negative consequence of technogenic contamination of the hydrosphere is rapidly reducing resources of fresh drinking water. The rapid deterioration of the quality of surface water is of particular concern. Water from many Russian rivers is not only unusable for drinking but is hardly good for swimming in it. Underground water is polluted by liquid factory waste discharged intensively and mostly without any control. Every year fewer and fewer sources of freshwater pass the standards. Not only freshwater, but also the ocean suffers technogenic chemical polluting. In Japan, the notorious Minamata disease was called after Minamata Bay where mercury-containing industrial wastewater was released by a chemical factory for many years. For the local population using mostly sea products for food this resulted in chronic Minamata disease.

Radioactive pollution

A separate problem that arose as late as in the last third of the 20th century is the problem of radioactive waste (Kovalev *et al.*, 1996).

To make the coverage of this problem sufficiently complete, at least a brief description of the general state and prospects of nuclear energy is required. In 1940, spontaneous nuclear fission of uranium was discovered; this reaction produces free neutrons and releases a very large amount of energy, which is much larger even than that of α-decay, let alone β-decay. The kinetic energy of repulsion of positively charged nucleus fragments is 50 million times higher than the energy of combustion of hydrogen atoms. The U235 fission-released energy is about 200 MeV, of which 86.1% immediately turns into heat, 5.6% gets scattered in the space with neutrinos, and 3.1% makes neutrons perform excitation and splitting of new nuclei. The remaining 5.2% appears as ionizing β- and γ-radiation, thus creating ecological problems associated with nuclear-power engineering. This ionizing radiation turns into heat as well, but far from instantly. If this part of energy could be "tamed" through its accelerated conversion into heat, then nuclear power would become not only environmentally friendly, but also twice as profitable, because nuclear fission products are isotopes of rare and noble metals. Totally, U235 fission yields more than 200 isotopes of 36 elements.

According to the International Atomic Energy Agency (IAEA), as of March 2000, 436 reactors of nuclear-power plants (NPP) with a total capacity of 351,718 MW worked in the world. The highest share of nuclear power used for electricity generation is reported for France (75%), Lithuania (73%), and Belgium (58%). In Russia it is less than 20%.

One ton of spent fuel from modern NPPs contains about 9 kg of nuclear fuel fragments and approximately as much plutonium. About 25% of the fragments are those completely stabilized, another 25% are solid fragments displaying low activity (about 1% of total activity), 16% are gaseous and light volatile products that add one more percent to the activity. The remaining 36% (over 3 kg) of the fragments give 98% radiation. Upon spent nuclear fuel discharging, the activity of nuclear fission products (NFP) amounts to 240–260 million curie (Ci) per ton. Let me remind you that Ci is the activity of 1 g of radium (equivalent to about $3.7 \cdot 10^{10}$ disintegrations per second) that is taken as the standard unit for measuring activity (another unit termed becquerel [Bq] equals one disintegration per second). The discharged spent fuel is kept underwater in a special pool for three years, after which its activity appears as low as 1 million Ci per ton due to disintegration of short-lived isotopes. But it takes five years of cooling before fuel assemblies (TVEL, Russian abbreviation for the "heat-releasing element," fuel rod) can undergo reprocessing. According to the technology currently used in Russia, three types of radioactive material are extracted from spent nuclear fuel: (1) elements that can be used as nuclear fuel (i.e., fissible uranium and plutonium); (2) fission fragments; and (3) transuranium elements. In turn, according to their activity, fission fragments are subdivided into low-, intermediate-, and high-level ones.

Before exhaustion of all known uranium and thorium reserves, NPPs are calculated to produce (even if using unimproved technologies) less than 0.1 km^3 of radioactive waste, of which only 3% is strongly hazardous. As compared to the continental crust volume, it is a vanishingly small value. The challenge is to store this waste reliably contained and prevented from coming into contact with the environment. High-level long-lived isotopes of cesium–strontium fractions and transuranium elements are of particular danger; the former require isolation for 500–600 years, whereas the latter must advisably be buried "for perpetuity." In Russia it was decided at the governmental level that all high-level radioactive waste must be solidified, but due solidification techniques were not specified. Many of the proposed and to some extent implemented techniques of storing radioactive waste (e.g., solidified in bitumen, concrete, technical stones, or vitrified in alumophosphate and borosilicate glass) do not reliably prevent radiation exposure to the environment. Self-heating bitumen and organic/mineral composite exposed to oxygen become oxidized and fire hazardous; concrete and technical stones are destroyed by ionizing radiation (radiolysis); upon contact with water, alumophosphate glass and borosilicate glass form acids that dissolve radionuclides. In some countries radioactive waste is contained in metal vessels and stored pending final disposal.

Equally geochemically unfounded is the project of constructing huge disposal facilities for high-level waste in permafrost on Novaya Zemlya (New Land). Specialists in geochemistry and geocryology have long provided evidence for intensive physicochemical processes that take place in frozen rocks due to the presence of the liquid aqueous phase in the form of thin films and concentrated solutions. Therefore, permafrost in no case can be regarded as an insulating matrix for chemically active and heat-releasing radioactive waste.

In solving the problem of preserving the tightly bound state of radioactive metals, which eliminates (or minimizes) the possibility of environmental releases, the leading role should undoubtedly be played by geochemistry that studies endogenous and exogenous chemical processes in control of terrestrial matter redistribution. Due to intensive development of military and civilian nuclear industries, radioactive waste disposal has become a global challenge to mankind (Duursma, 1996). As a conceptual solution of this problem, geochemists propose constructing geological systems that are close to natural ones, high-stable, and based on aluminosilicate rocks; these could provide an equilibrium state of waste for a long time (Kovalev *et al.*, 1996). The most reliable way to manage uranium is the sorption geochemical barrier.

In the United States in the Nevada desert, a special mine to be employed as a deep geological repository storage facility, the Yucca Mountain Nuclear Waste Repository, has been built within Yucca Mountain. As a mandatory safety condition of this facility, its design stipulated that the groundwater would not reach the lowest horizons of the mine in the next 10,000 years. However, an international group of experts, including a scientist from Novosibirsk, denied the possibility that this condition can be met because there was evidence for such a happening in the previous period of 10,000 years. Notwithstanding communication of this expertise to the U.S. authorities and its publication in the IAEA bulletin, the Americans are already using this mine as a radioactive waste repository.

Another problem of radioactive waste management is disposal of spent fuel of atomic-powered vessels. Currently, it is simply discharged overboard into the ocean depths. Each nuclear state has its own radioactive waste "pits." Naturally, these are monitored; thus far, no negative consequences have been registered.

Biological pollution

This issue is more of biological than geochemical nature, but the interests of biology and geochemistry intersect where biological migration of chemical elements is concerned. Changes in the biota species composition are accompanied by biogeochemical changes in the area in question. And in some cases, invasion of an alien species leads to serious environmental problems. Let us consider a concrete example.

There exists a most harmful water plant called *Elodea canadensis*. Its harmfulness is dual. When it blossoms (in the warmest season of the year), it fills the entire water layer from bottom to surface so that even fish cannot get through it. In the 2.5 m deep Chivyrkuysky Bay of the Baikal people have to make special tunnels in these thickets. But this is not the worst yet: in autumn, *Elodea*'s decay is accompanied by formation of water-contaminating toxins, thereby turning the biological pollution into a chemical one.

Elodea spreads from west to east; it has already reached Transbaikalia (i.e., the Amur watershed). So far, no effective methods to control this pest have been invented, and its expansion can turn into a planetary disaster.

Because the cause of alien species invasion is often of anthropogenic nature, it may serve as a sign of anthropogenic pollution of the ecosystem. For example, a voracious fish, the Amur sleeper (*Percottus glehni*), previously found exclusively in the lower reach of the Amur River unexpectedly appeared in its upper reach, the Argun River. As it is now widely known, "cooperative efforts" of China and Russia have made the Amur River very dirty, thus forcing this fish to look for cleaner water.

Let's not get absorbed in this subject; here we just say that in recent years the invasion problems have already acquired a mass character and become critical.

Ecological and geochemical monitoring

Ecological (environmental) monitoring is a comprehensive monitoring of the current state of the environment, including components of the natural environment and natural ecological systems with processes and phenomena occurring there; also, it includes assessment and forecast of changes in this state. A mandatory condition for the monitoring is continuous series of observations and a uniform methodology for different services.

Lecture 9
Geochemistry of systems

The term *system* comes from Greek and means "whole concept made of several parts" (literally "composition"). It can be defined as a regularly interacting group of items forming a unified whole (Dictionary of Foreign Words, State Publishing House of Foreign and National Dictionaries, Moscow, 1955 [in Russian]). In natural sciences, a system is understood as a certain set of natural components that has the property of integrity due to regular direct interactions and feedbacks between its members. Specificity of every system (i.e., its difference from other systems), consists in quality and quantity of its members and the type of their arrangement.

The systematic approach in natural science originates from the teachings of ancient Greek philosophers. In particular, Aristotle (384–322 BCE) regarded the world as a single dynamic system consisting of four elements: earth, water, air (i.e., solid, liquid, and gaseous matter), and fire (energy). With time and distance added, this gives major parameters of the material world that we still use when describing static and dynamic systems. The notion that matter has an atomic structure also arose in ancient Greece about 2500 years ago.

The importance of a systematic approach in science, and specifically in geochemistry, was emphasized by Vernadsky who thought it impossible to divide natural phenomena into independent parts without doing harm to the resultant conclusion.

In the middle of the 20th century the development of a general theory of systems pushed this issue to a quantitatively new and higher level. Each system, if it has the right to be called so, has special properties underlain by interactions of its constituents (the emergence property), which is amply exemplified at any level from micro to macro. Therefore, it is hardly possible to predict the properties of a large system from characteristics of its separate members.

The basic aspects of the system approach to research are as follows:

- The hierarchy of systems (from micro to macro).
- The system structure (i.e., the totality of its components and modes of their interrelationship, including direct interactions and feedbacks, both positive and negative).
- Identification of the major members of the system.
- Identification of the major conflicting sides among the system constituents, whose interactions underlie the system functioning and essence, for a system is not just a unity but a unity of opposites that provides a dynamic equilibrium required to prevent its "getting off-scale."

The Earth system includes a number of subsystems; these are the Earth's core, the mantle, the crust, the biosphere, the ocean, and the atmosphere. All of them interact with one another maintaining a certain dynamic equilibrium. In the event that this equilibrium is disturbed, the system tries to regain it using mechanisms developed in the course of evolution; until now, it was a success. One of the basic laws of geology is seen to be a profound interrelationship involving life development on Earth, sedimentation, tectonic phenomena, and magmatism. In other words, the features of an earlier sedimentation should be reflected in geochemistry of geological bodies formed later.

The modern concepts of natural science, which are rather numerous, describe a generalized model of interaction of the Earth's subsystems. However, in nature, there occur some fluctuations caused by fortuitous events. A fortuity happens due to a multiplicity of factors and the probabilistic character of interactions in the system. Therefore, our ideal models can hardly be expected to adequately represent the entity under study. It will only be an approximation to the reality.

Stability of geosystems obeys the Wiener-Shannon-Ashby's law of requisite variety, which states that any cybernetic system appears stable enough to block internal and external disturbances only if it has a sufficient requisite variety itself. Then in case of failure of its constituent a suitable substitute can possibly be found. Hence, a prerequisite to stable existence of a system is maintenance of its requisite variety. This is quite reasonable for the biosphere, which is a living system with many of the variety-maintaining mechanisms mentioned above. But what about inert geosystems? Do they possess any mechanisms for maintaining their stability? When assuming that inert geosystems are capable of self-developing, we have to assume that they are capable of preserving their stability. In turn, system stability is controlled by thermodynamics-based laws of self-development; specifically, nonequilibrium thermodynamics implies the tendency of a system to the stable state, which actually is stability. Thus, Prigozhin's laws of nonequilibrium thermodynamics "contribute" to maintaining the stability of inert geosystems.

The law of the environment-supported development (existence) of a natural system states that such a system can develop (and exist) only by using material, energy, and information coming from its environment. Self-development of an isolated system is impossible. It follows from this law that (1) nonwaste production is unattainable in principle; (2) a highly organized system is a potential threat to the low-organized one; (3) the development of the Earth's biosphere is supported not only by internal resources of the planet but also by the space systems (primarily solar).

As a system, the biosphere consists of four major components: the lithosphere, the hydrosphere, the atmosphere, and living matter. The latter plays the leading role. The center of the biosphere is the phytosphere, where photosynthesis provides accumulation of solar energy. The key elements of the modern phytosphere are forest landscapes of the dry land and upper horizons of the ocean. The landscape is the main structural element of the surface biosphere; it includes soil, weathering crust, living organisms, surface and groundwater, silts, and surface atmosphere. The landscape consists of bioinert bodies, and it is a bioinert system of a higher order itself (Perelman, 1987). Unlike its components such as phase or hole, it has a requisite variety sufficient for its self-development. The law of geochemical landscape by Alekseenko (Alekseenko, Alekseenko, 2002) states that the changes that have occurred in a certain part (tier) of the landscape will affect practically all parts of this landscape due to their interconnections. This underlies the formation of cause–effect chains that necessarily should be known for making geoecological forecasts.

An important element of the biosphere is soil. According to Dokuchaev, soil is an independent natural body having its own history of development; it is a result and a function of the common effect of local climate, vegetation, living organisms, relief and age, as well as its orthometric height and parent material (i.e., bedrock). Soil is obviously the basis of the entire plant world.

The ocean is a system that, due to its internal mobility, serves as a kind of buffer in various geochemical cycles. In accordance with Henry's law constants, the dynamic equilibrium between the ocean and the atmosphere regulates the content of two important constituents of the atmosphere – oxygen and carbon dioxide. The ocean plays a crucial role in the carbon cycle; besides, it is an important climatic factor.

The Earth's crust as a system

Complex natural systems of the exosphere are a result of the common activity of the atmosphere, the hydrosphere, the lithosphere, and the biosphere. The Earth's crust receives energy from two sources: from the cosmos (mainly from the Sun) and from the Earth's interior. When merged in the crust, these two oppositely directed flows create a complex thermodynamic situation.

The Earth's crust, as we already know, consists of several layers, that is, the basalt layer, the granite layer, and the biosphere. The two lower layers form a boundary between the mantle and the biosphere; apart from other their functions, they act as a transit zone for supplying food to habitat of living organisms and discarding their metabolic products. For this purpose nature uses a variety of processes, mostly those involving water.

Specificity of disperse exogenous systems

In hypergenic (near-surface) conditions, many geochemical systems are characterized by high dispersity of the solid phase. These include clay rocks of weathering crusts, soils, silt, and bottom sediments. Their most important feature is a large specific surface of mineral particles, which predetermines the role of surface phenomena in geochemical processes.

For example, the effective properties of particles of clay size in their interaction with other phases (like aqueous solution) are dictated by characteristics of their surface rather than volume.

The surface of the solid phase has excess energy, which, upon contact of a mineral particle with the aqueous phase, causes an extremely thin water film to form on the mineral surface. In dispersed (as well as porous) rocks and soils the following forms of moisture are distinguished:

- Bound water including lattice (crystallization) water and adsorbed moisture
- Transition-type water, including osmotic and capillary moisture
- Free (gravity) water

Together, adsorbed and osmotic water form a 20 nm thick water layer. Its appearance is caused by a structural rearrangement of water molecules using solid phase surface energy and by appropriate changes in their physicochemical properties. The layer immediately surrounding the mineral surface is called the double electric layer; it is formed by chemically adsorbed moisture where H^+ is chemically bonded to surface radicals of the particle, and hydroxyl ions are attracted by positive charges of the formed complexes. The thickness of the layer of chemisorbed moisture is about 1 nm. It should be noted that this layer is not continuous and depends on the density of its constituent uncompensated bonds on the mineral surface. These are active sites, positive or negative, and usually unequal in number. The majority of rock-forming minerals have mostly negative active sites on the surface.

Aqueous films play a significant and often decisive role in dissolution of mineral phases, heat, and mass transfer in disperse systems, disintegration of rocks and ores, and chemical weathering in the cryogenesis zone; they change the strength and rheological characteristics of rocks and the speed of seismic waves. While in rocks with a small specific surface the amount of structured moisture is vanishingly small in comparison with the mass of the rock, it appears a notable contribution to the total mass balance in highly dispersed systems. Let us consider properties and geochemical behavior of aqueous films.

Adsorbed bound moisture comprises water that is chemically bonded to active sites on the mineral surface with a binding energy of 40–120 kJ mol^{-1} and water bound to the solid surface with weaker van der Waals physical bonds (binding energy < 40 kJ mol^{-1}). Water adsorbed in the osmotic manner results from selective diffusion of water molecules toward the particle surface ("surface" osmosis) due to the concentration difference between the film-forming solution and the free aqueous phase. Because of osmotic pressure, this water produces a wedging effect which leads to rock swells. The surface osmosis is limited by the total moisture content of the system and concentration of the solution.

It is evident from the above that in nature there is water of various types with substantial difference in its properties. The structural multiplicity of moisture, as one of the most important components of the environment, determines the diversity and complexity of relationships in water–rock and water–organic matter systems. Accordingly, H_2O cannot be unconditionally regarded as an independent component, like in case of construction of solubility diagrams, etc.

An important conclusion is that phase transitions of H_2O are not properly shown in the commonly used diagram because ice, liquid, and steam may disagree with H_2O stoichiometry. Consequently, the phase transitions of H_2O can occur not at a certain temperature (with fixed pressure), as it is commonly believed, but within some temperature range. This issue has been virtually unaddressed yet, though this suggestion has a real basis.

As to its properties, moisture in the form of a film differs much from free water. Under the effect of mineral surface energy, the adsorbed moisture acquires a structure, thus diminishing its chemical potential and changing its other parameters, such as the boiling and freezing points, density, etc. The dielectric constant of water also changes (in physically bound water it reaches 40–50, while in chemically bound water it is 4–6), followed by all other constants associated with electrolytic dissociation. Accordingly, the water's dissolving capacity changes, too. At a temperature below zero, the dielectric properties of water undergo additional alteration. Thus, disperse cryogenic systems are specific in both chemical and thermodynamic terms. The thermodynamic status of aqueous films adsorbed on the mineral surface

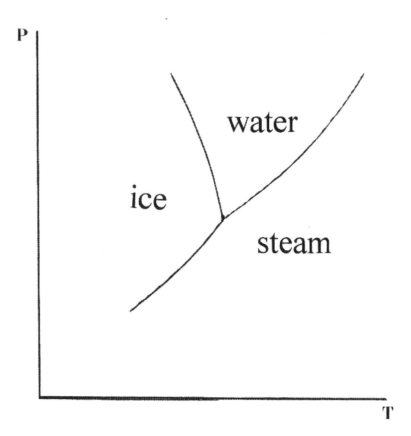

Figure 9.1 Phase diagram of H_2O.

is still debatable (see, e.g., Storonkin, 1967). So, both thermodynamics and chemistry of aqueous films require special consideration. Thermodynamics of water films is in the focus of studies by Rusanov (1967) and others. Thermodynamics of phase relations in the system "solid minerals–water films" is discussed in the chemical (Lopatkin, 1987) and geochemical (Urusov *et al.*, 1997; Ptitsyn, 1998) literature. Specifically, the debatable issue is the location of aqueous films in the phase diagram (Storonkin, 1967). They do not form a separate phase because they exist exclusively on a mineral substrate. This was the reason for the term "nonautonomous phase" proposed by Urusov and colleagues in 1997. Figure 9.2 illustrates the location of this nonautonomous phase in the phase diagram (Ptitsyn *et al.*, 2009).

The stability area of the nonautonomous phase adsorbed on the mineral surface is limited by the complex surface A-B-C-F-G-E-H-D-S-R and the

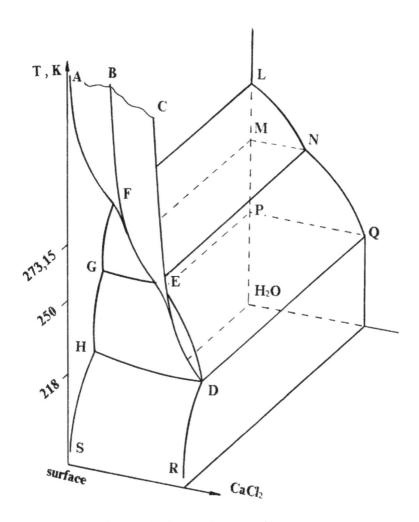

Figure 9.2 Phase diagram with the nonautonomous phase.

coordinate planes. Aqueous films mediate the "water–rock" interaction, and it is their properties that determine the character of this interaction. Therefore, water films are not just the Nernst's diffusion layer but a certain boundary, albeit thin, having special chemical properties. In aqueous films and electrolyte solutions the chemical potential of water decreases in a similar way (Fig. 9.3). Therefore, the surface of mineral particles competes with ions of dissolved substances for water dipoles.

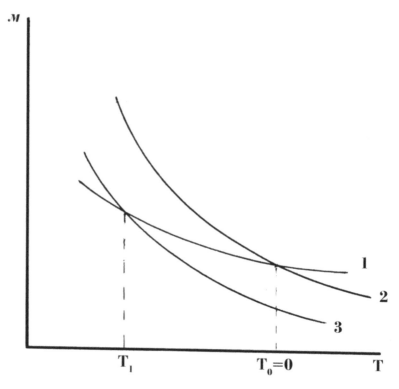

Figure 9.3 Temperature changes in the H_2O chemical potential: 1, in ice; 2, in ordinary
bulk water; 3, in bound (either to the mineral surface or ions in solution)
water.

The proportion of bound water dipoles increases with increasing concentration of the solution, thereby thinning the adsorbed water films and
gradually opening the direct access of the bulk solution to the mineral surface, which, in turn, changes the kinetics of heterogeneous reactions. In
cryogenic conditions, the water–rock systems have their own specificities.

Cryogenic systems

The state and behavior of geochemical systems at a temperature below zero
(Celsius) are fundamentally different from those at positive temperatures.
This results from ice crystallization and its peculiarities. The principal difference lies in the fact that at $t < 0°C$ the kinetics of reactions in the water–
rock system becomes volume dependent instead of concentration dependent

because the concentration of the aqueous solution is buffered due to crystallization of the "excess" ice (Fig. 9.4).

Thus, at constant low temperatures, the concentration of a reagent will remain constant (and very high) regardless of its consumption, while the volume will decrease.

Because all natural systems are open, they cannot be strictly described by the equilibrium thermodynamics machinery. But in some cases we can talk about local temporary equilibrium. Besides, we can talk about stepwise equilibrium where methods of equilibrium thermodynamics can be used not only to analyze the results, but also to study a process, as is currently practiced in programming of physicochemical models.

In conclusion, we will briefly review technogenic geochemical systems, including point, linear, and areal ones. The boundaries between these systems are special zones with special geochemical flows.

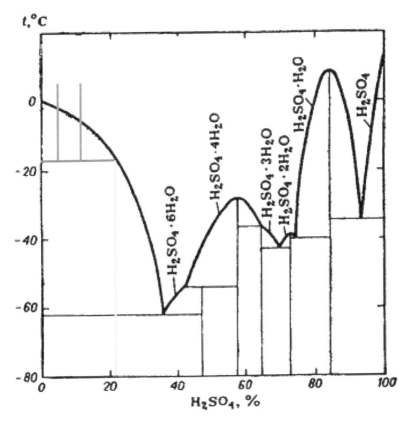

Figure 9.4 The effect of cryo-buffer.

Conditionally, point technogenic systems are large cities. They are characterized by a variety of activities, and hence by a complex pollution of the environment. They import energy, various raw materials and products and export their produce. Cities discharge into the environment their industrial and domestic waste, which comes to the atmosphere and the hydrosphere. Northern cities having cold winters discharge pollutants in one spring volley together with accumulated snow. Also, a large city affects the geological environment with its large weight, which leads to its gradual sinking. In the world there are cities that during their lifetime have sunk by 5–7 m. A city is the source of thermal pollution, and temperature in the city is usually higher than in the suburbs. Depending on the size of the city, the contaminated zone beyond its borders ranges from 5–10 km.

Linear technogenic systems include roads, railways, and pipelines. The contaminated territory on each side of the road is approximately 10–15 m wide. Pipelines are a chemically hazardous entity only in case of leaks.

Areal technogenic systems are agricultural territories. Their pollutants are of a special type, like fertilizers, pesticides, fuels and lubricants, metal fragments of worn out tools, animal disinfectants.

Lecture 10
A generalized model of Earth as a self-developing system

So, Earth as a planet was formed from cosmic matter in the course of formation of the solar system. Its composition and structure obey the law of matter distribution in the solar system; that is, the specific gravity of the planets decreases with increasing distance to the Sun. The same law is valid for the Earth's geospheres where light elements are moved to the periphery. Further evolution of Earth was determined by common action of external factors, internal processes, and laws of self-development. Of the external permanent factors, the most important are energy ones, that is, solar energy and its alteration with time, as well as gravitational energy that varies cyclically depending on the location of celestial bodies relative to Earth. Meteorite shower, comet effects, solar wind, etc. can be assigned to the category of secondary factors.

In accordance with the laws of thermodynamics, Earth as the member of a larger (cosmic) system has adapted to the cyclic nature of cosmic "life" and subordinated its internal processes to it. This explains the correlation of terrestrial tectonic and magmatic processes, as well as the global geochemical cycle, with the galactic year.

Another important factor in the Earth's life and evolution is living matter, which originated and constructed the biosphere and its system of matter and energy cycling that is now inherent to the planet as a whole. Together, external and internal factors provide interconsistency of direct interactions and feedbacks that maintain the dynamic equilibrium of this most complex and "whimsical" organism where deterministic dependencies intertwine with stochastic dependencies, thus allowing occurrence of illogical chaotic events. These additional degrees of freedom reduce the reliability of our models, and hence forecasts.

Another important factor for the Earth's evolution is energy of radioactive decay of natural radionuclides (i.e., uranium, thorium, and potassium) unevenly distributed on the planet due natural geochemical processes whose alteration with time is an inherent feature of the evolution. Again, this exemplifies twisted cause–effect relations (direct and reverse) in the most complicated Earth system.

Another "capricious" participant of biospheric and geochemical life is the Earth's climate. Naturally, it is strongly influenced by cosmic factors, first of all by the Sun, and also by comets that, upon approaching Earth, can cause a cold snap. Fluctuations of gravitational energy coming to Earth also affect its climate. Cyclic fluctuations of magmatic activity caused by these gravitational waves provide recurrence of intensive volcanism. Volcanoes are the major suppliers of carbon dioxide to the atmosphere. Carbon dioxide is responsible for the greenhouse effect, and hence for temperature on the planet (as well as productivity of the biosphere). All this has a direct correlation with global climate changes. However, lately, people more and more often talk about unevenness and a mixed character of the Earth's climatic changes, which are frequently even opposite in sign. What is the reason for this phenomenon? An unequivocal answer has not been found yet. Nevertheless, some suppositions can be made.

The uneven distribution of natural radionuclides in rocks leads to unequal "heat supply" to planet regions. Figuratively speaking, one house/region has a warm floor comfortable for its residents, while another one is devoid of heating, which results in cold feet and frequent chills. Another example: in China there is a territory (Liaoning Province) where the ground is warm in winter (up to +17°C) and cold in summer (reaching −12°C), thus serving as a refrigerator for the residents. What causes this phenomenon? The scientists found an explanation. "The region in question is underlain by a tectonic fault along which, as regular as clockwork, compression occurs in winter and expansion of the rocks in summer" (Bgatov, 1993, p. 80). This is why in winter the ground temperature is as high as in the subtropical climatic zone, while in summer the cold snap results from the throttling effect of carbon dioxide coming from the depth.

To sort out the diversities of climate changes on the planet, first of all we need to draw a map of heat flows using many thousands of reported measured data. In case of blank spots on this map (that most probably will occur) additional measurements should necessarily be taken. These studies will not only contribute to fundamental knowledge but also may be of a practical use in various sectors of the national economy, primarily in agriculture. Such a project cannot be realized by specialists of one country; it requires a special international program.

Let us return to the Earth's geochemical evolution. To deal with these multifactor jungles, one has to pluck up his/her spirits and neglect "trifles." This is not easy because when rejecting trifles one can "throw out the child along with the bath" (from Martin Luther's "*Das Kind mit dem Bade ausschutten*"). OK, my reader, let's set aside the doubts and go ahead.

First of all, let us paint with large strokes a picture of factors involved in geochemical processes within the Earth. Intensive parameters vary within widest ranges: temperature – from negative values (−80°C and lower) to about 2000 °C to 3000 °C in the Earth's center; pressure – from Pa fragments in the atmosphere to about 3 million atmospheres in the central parts of the planet. Extensive parameters (which are much more numerous) depend on the system's composition and "behavior." Perhaps, these should be assigned to characteristics of the system rather than parameters because they are the system's "produce." This seditious statement leads to the necessity to distinguish between "intensive" and "extensive" parameters. Intensive parameters "by definition" do not depend on the system's composition; they are set from the outside. But what sets them? An external system where "our" system is a constituent. If any parameter considered to be "extensive" for our system has been buffered by the external system, it becomes intensive for our system. Thus, intensive parameters are those whose values are set (buffered) by the external system. For example, temperature commonly thought to be an intensive parameter may become extensive if it varies depending on the system's composition, like in case of conditions favorable for exothermic reactions. In contrast, the oxidation potential or even concentration (activity) of some component (commonly believed to be extensive parameters) can be buffered from the outside and then be ranked as intensive ones.

Consequently, geothermal and geobaric gradients are just "the average fever heat for the hospital," while the real situation in the Earth's depths is much more complicated. Then, are these planet-averaged parameters at all useable? It depends on the scope of problems we are solving. For global problems – yes, they are. To solve local problems, we need other, more detailed raw data. Thus, in studies of systems and processes there involved, the scope of the tasks to be solved is of great importance.

Here, it seems appropriate to formulate the challenge in geochemistry that has not been explicitly stated yet: there is a long felt need to develop a general system of interrelated geochemical processes. If, of course, such a challenge can be regarded by the scientific community as a feasible one.

Now let us talk about the "human factor." I ask my students the following question: shall we regard the emergence of humans on the planet as a natural (and hence, necessary) product of the biosphere's evolution, or is it a sore, an abscess on the biosphere's body caused by an unfavorable combination of external factors? The answer to this question underlies the concept of nature management. What is to be done? Should we leave everything "as is" and let nature itself deal with its problems or must the biosphere receive an intensive treatment? If the latter is the case, we will have to undertake self-treatment which is fraught with errors.

The problem is that humans are the only biological species who, on their own volition or whim, choose the pathway of their development. Having turned into *Homo sapiens*, people ceased to obey the laws of the animal kingdom (although implicitly animal instincts are still preserved) and began to act wilfully. As a result, according to Vernadsky, by the beginning of the 20th century the anthropogenic impact has reached the scale of geological processes. Accordingly, human intervention in geochemical processes has increased, thus causing the appearance of the anthropogenic geochemical factor that is beyond any strict scientific forecast.

Let us itemize the most significant "achievements" of humans in geochemical life of the planet.

1   Reduction of the forest area. At first, at the dawn of mankind, separate forests were burned down only in order to drive out the game. Then new lands were needed for pasture and farming. As a result, the growing human population was entailed by growing intensity of desertification (a vivid example is the Sahara). Another result consisted in gradually misbalancing amounts of oxygen and carbon dioxide in the atmosphere (since forests are the lungs of the planet), which could not but cause climate changes with all the ensuing consequences.

2   Plowing. At first, when humans were not numerous and land was in abundance, plowed fields that have become barren were abandoned by plowmen who simply moved elsewhere. Having lost its fertile properties, the used land was exposed to wind and water erosion, thus turning into a desert. In today's world, in open areas of the United States, Kazakhstan, etc. plowed lands are defenseless against the onset of strong winds that literally "cut off" the fertile layer as thick as 7–10 cm. Currently, about 1.5 billion hectares are involved in the world agriculture. Every year, about 8 million hectares (on average) are redirected for other economic needs, and more than 7 million hectares happen to be withdrawn from agriculture due to various types of soil degradation (e.g., in 1934 in the United States a dust storm destroyed 45 million hectares of fertile land). By promoting soil degradation in various ways (unwise farming, overgrazing, chemical pollution, acid rain), people deprive themselves of the future. Soil, as the basis of biosphere's life, is the "ecological shield" of Earth.

3   Pollution of surface and groundwater. Freshwater is already deficient on the planet. Importantly, not only people and animals suffer from its shortage, but also all freshwater hydrobionts. Some geochemical barriers (the major one is the sorptive barrier) are responsible for removal pollutants from natural waters, but they cannot cope with this work any longer.

4   Acid precipitation. By sulfur dioxide and nitrogen oxide emissions to the atmosphere people promote sulfuric and nitric acid precipitation; its free circulation around the planet causes degradation of forests, soils, buildings, and historic landmarks; also, it intensifies mobility of chemical elements (including toxic ones).

5   Fuel combustion. Hydrocarbon raw materials accumulated within the planet during many millions of years are being burned by people at an incredible rate. This is accompanied by emission of a huge amount of carbon dioxide which Earth fails to "digest." Due to the greenhouse effect, this possibly will result in global warming with all the ensuing consequences.

6   Contamination of the land surface with substances alien to the biosphere (mostly plastics) which nature can hardly deal with. Pollution of the ocean surface with oil products, which leads to mass mortality of marine hydrobionts; the possible consequences of this impact have not been evaluated yet.

7   Radioactive contamination of the biosphere. It has been shown that radiation affects the genetic code of organisms by changing the structure of chromosomes. Thus, the consequences of atomic explosions affect next generations. For example, in the Republic of Altai accessible for winds from Kazakhstan people faced the "yellow children" disease that manifested itself many years after explosions at the Semipalatinsk testing area.

Thus, now geochemistry has a social component that further complicates its methodology and reduces the reliability of scientific forecasts.

References

Abyzov, S.S., *et al.* (2002) *Bacterial Paleontology.* Publ. House PIN RAS, Moscow, 188 p. [in Russian].

Alekseenko, V.A. (2000) *Environmental Geochemistry.* Logos, Moscow, 626 p. [in Russian].

Alekseenko, V.A. & Alekseenko, L.P. (2002) *Biosphere and Vital Functions: A Tutorial.* Logos, Moscow, 212 p. [in Russian].

Alekseenko, V.A. & Alekseenko, L.P. (2003) *Geochemical Barriers: A Schoolbook.* Logos, Moscow, 144 p. [in Russian].

Ayvazyan, S.M. (1967) *Convergence Series of Valence Isotopes.* Publ. House of the Academy of Sciences of the Armenian SSR, Yerevan, 318 p. [in Russian].

Barabanov, V.F. (1985) *Geochemistry: A Textbook for Universities,* Nedra, Leningrad, 423 p. [in Russian].

Barnes, H. & Czamanski, G. (1970) Solubility and migration of ore minerals. In: *Geochemistry of Ore Deposits.* Mir, Moscow [in Russian].

Battistuzzi, F.U., Feijao, A., & Hedges, S.B. (2004) A genomic timescale of prokaryote evolution: insights into the origin of methanogenesis, phototrophy, and the colonization of land. *BMC Evolutionary Biology,* 4, 44.

Betekhtin, A.G. (1953) Hydrothermal solutions, their nature and processes of ore formation. In: *Basic Issues of the Doctrine of Magmatogene Ore Deposits.* Moscow [in Russian].

Bgatov, V.I. (1993) *Approaches to Ecogeology.* Publ. House of Novosibirsk Univ., Novosibirsk, 222 p. [in Russian].

Bjerrum, I. (1961) *Metal Amines Formation in Aqueous Solution.* IL, Moscow [in Russian].

Bulkin, G.A. (1972) *Introduction to Statistical Geochemistry.* Nedra, Leningrad, 208 p. [in Russian].

Clark, S.P. (Junior) (1961). State equations and polymorphism at high pressure. In: *Reserches in Geochemistry.* Transl. from English. IL, Moscow, pp. 600–618 [in Russian].

Day, M.C. & Selbin, J. (1969) *Theoretical Inorganic Chemistry.* Transl. from English. Khimiya, Moscow, 432 p [in Russian].

Dobretsov, N.L., Kirdyashkin, A.G. & Kirdyashkin, A.A. (2001) *Abyssal Geodynamics*. Publ. House of SB RAS, "Geo", Novosibirsk, 409 p. [in Russian].

Dubnischeva, T.Ya. (1997) *Concepts of Modern Natural Science: A Textbook*. UKEA, Novosibirsk, 832 p. [in Russian].

Duursma E.K. (1996) *Carroll JoLynn Environmental Compartments. Equilibria and Assessment of Processes Between Air, Water, Sediments and Biota*. Springer-Verlag, 280 p.

Fedoseeva V.I. (2003) *Physico-chemical Regularities of Migration of Chemical Elements in Frozen Soils and Snow*. Yakutsk, Izd-vo IMZ SB RAS, 138 p. [in Russian].

Fersman, A.E. (1959) *Selected Works in 4 Volumes*. Publ. House of AS USSR, Moscow, vol. 4, 858 p. [in Russian].

Garrels, R.M. & Christ, C.L. (1968) *Solutions, Minerals, Equilibriums*. Mir, Moscow, 368 p. [in Russian].

Glansdorf, P. & Prigogine, I. (1973) *Thermodynamic Theory of Structure, Stability and Fluctuations*. Mir, Moscow, 280 p. [in Russian].

Goldschmidt, V.M. (1954) *Geochemistry*. Clarendon Press, Oxford, 730 p.

Golubev, V.S. & Shapovalova, N.S. (1995) *Man in the Biosphere: A Schoolbook*. Varyag, Moscow, 128 p. [in Russian].

Haase, R. (1967) *Thermodynamics of Irreversible Processes* (*Thermodynamik der irreversiblen prozesse*). Transl. from the German. Mir, Moscow, 544 p. [in Russian].

Hawking, S. (2015) *The World in a Nut Shell*. Transl. from English. Amphora, SPb., 218 p. [in Russian].

Helgeson, H.C. (1969) Thermodynamics of hydrothermal systems at elevated temperatures. *American Journal of Science*, 267(7).

Hentov, V.Ya. (2005) *Environmental Chemistry: A Textbook for Engineering Faculties*. Phoenix, Rostov-on-Don, 144 p. [in Russian].

Ivanov, A.B. (1996) *The Biosphere Science*. Khabarovsk – Komsomolsk-on-Amur, 48 p. [in Russian].

Kapustinsky, A.F. (1956) On theory of Earth. In: *Problems of Mineralogy and Geochemistry*. Publ. House of AS USSR, Moscow [in Russian].

Karapet'yants, M.H. (1949) *Chemical Thermodynamics*. State Publ. House for Chemical Literature, Moscow – Leningrad, 546 p. [in Russian].

Knorre, D.G., Krylova, L.F. & Muzykantov, V.S. (1990) *Physical Chemistry: A Textbook for Biological Faculties*. Higher School Publ. House, Moscow, 416 p. [in Russian].

Kolonin, G.R. & Ptitsyn, A.B. (1974) *Thermodynamic Analysis Conditions of Hydrothermal Ore Formation*. Nauka, Novosibirsk, 102 p. [in Russian].

Kovalev, V.P., Mel'gunov, S.V., Puzankov, Yu.M. & Raevsky, V.P. (1996) *Prevention of Uncontrolled Spread of Radionuclides in the Environment*. Publ. House of SB RAS, NITS OIGGM, Novosibirsk, 162 p. [in Russian].

Kumar, S. (2005) Molecular clocks: four decades of evolution. *Nature Reviews Genetics*, 6 (8), 654–62.

Lopatkin, A.A. (1987) New trends in thermodynamics of adsorption on solid surfaces. In: *Physical Chemistry: Modern Problems*. Khimiya, Moscow, pp. 89–127 [in Russian].

Meadows, D., *et al.* (1972) *Limits to Growth*. Moscow, IKTS "Akademkniga" [in Russian].

Milne, D., *et al.* (1985) *The Evolution of Complex and Higher Organisms*. NASA SP-478, Washington, DC, pp. 14–15.

Parmon, V.N. (1998) *Introduction to Thermodynamics of Nonequilibrium Processes: A Textbook for Students of Chemistry and Ecology Departments*. Publ. House of Novosibirsk Univ., Novosibirsk, part 1, 106 p.; part 2, 95 p. [in Russian].

Perelman, A.I. (1987) *Studying Geochemistry*. Nauka, Moscow, 152 p. [in Russian].

Perelman, A.I. (1989) *Geochemistry*. 2nd ed. Higher School Publ. House, Moscow [in Russian].

Prigogine, I. (1960) *Introduction to Thermodynamics of Irreversible Processes*. IL, Moscow, 150 p. [in Russian].

Ptitsyn A.B. (1998) The special properties of film solutions and their role in geochemical processes. *Geochemistry*, 12, 1291–7.

Ptitsyn, A.B. (2006) *Theoretical Geochemistry*. Acad. Publ. House "GEO", Novosibirsk, 180 p. [in Russian].

Ptitsyn, A.B. (2013) *Geochemistry of the Biosphere*. Novosib. State Univ., Novosibirsk, 238 p. [in Russian].

Ptitsyn, A.B., Abramova, V.A., Markovich, T.I. & Epova, E.S. (2009) *Geochemistry of Cryogenic Oxidation Zones*. Nauka, Novosibirsk, 88 p. [in Russian].

Purtov, P.A. (2000) *Introduction to Nonequilibrium Chemical Thermodynamics*. Novosib. State Univ., Novosibirsk, 97 p. [in Russian].

Ringwood, A.E. (1982) *The Origin of Earth and Moon*. Nedra, Moscow, 293 p. [in Russian].

Roslyakov, N.A., *et al.* (1996) *Ecogeochemistry of the West Siberia. Heavy Metals and Radionuclear Substances*, Scientific editor G.V. Polyakov. Publ. House of SB RAS, SPC UIGGM, Novosibirsk, 246 p. [in Russian].

Rundkvist, D.V. (1965) The one about the General regularities of the geological development. *L. VSEGEI*, 1, 79–90.

Rusanov, A.I. (1967) *Phase Equilibriums and Surface Phenomena*. Khimiya, Leningrad, 388 p. [in Russian].

Scherbov, B.L., Lazareva, E.V. & Zhurkova, I.S. (2015) *Forest Fires and Their Consequences*. Acad. Publ. House "GEO", Novosibirsk, 154 p. [in Russian].

Selinov, I.P. (1990) *Structure and Systematization of Atomic Nuclei*. Nauka, Moscow, 112 p. [in Russian].

Sharkov, E.V. & Bogatikov, O.A. (2001) Early stages of tectono-magmatic development of Earth and Moon: similarities and differences. *Petrology*, 9(2), 115–139 [in Russian].

Shvartsev, S.L. (1998) *Hydrogeochemistry of the Hypergenesis Zone*. Nedra, Moscow, 366 p. [in Russian].

Smith, F.G. (1968) *Physical Geochemistry*. Transl. from English. Nedra, Moscow, 475 p. [in Russian].

Solov'ev, V.A. & Solov'eva, L.P. (2013) *The Global Ecology (Ecology of the Earth's Geospheres): A Textbook*. Kuban State Univ., Krasnodar, 465 p. [in Russian].

Solov'eva, L.P. (2013) *Geochemistry Fundamentals: A Textbook*. Kuban State Univ., Krasnodar, 297 p. [in Russian].

Starostin, E.G. & Lebedev, M.P. (2014) Properties of bound water in disperse rocks. Part I. Viscosity, dielectric permeability, density, heat capacity, surface tension. *Kriosfera zemli*, XVIII(3), 46–54 [in Russian].

Storonkin, A.V. (1967) *Thermodynamics of Heterogeneous Systems*. Leningrad Univ. Press, Leningrad, 443 p. [in Russian].

Taysaev, T.T. (2007) The phenomenon of cryobiogeneses and geochemical self-organization of permafrost landscapes. *Vestnik RAEN*, (2) [in Russian].

Ugai, Y.A. (2002) *General and Inorganic Chemistry*. Higher School Publ. House, Moscow, 527 p. [in Russian].

Urey, H.C. (1947) The thermodynamic properties of isotopic substances. *Journal of the Chemical Society*, 562–581 [in Russian].

Urusov, V.S., Tyson, V.L. & Akimov, V.V. (1997) *Geochemistry of Solids*. GEOS, Moscow, 500 p. [in Russian].

Uyeda, S. (1980) *The New View of the Earth*. Transl. from English. Mir, Moscow, 213 p. [in Russian].

Vakhromeev, G.S. (1995) *Environmental Geophysics*. ISTU, Irkutsk, 216 p. [in Russian].

Verhoogen, J., Turner, F.J., Weiss, L.E., Wahrhaftig, C. & Fyfe, W.S. (1974) *The Earth: An Introduction to Physical Geology*. Transl. from English. Mir, Moscow, 874 p. [in Russian].

Vernadsky, V.I. (1954) Essays on geochemistry. In: *Selected Works*. Publ. House of AS USSR, Moscow, Volume 1, pp. 5–392 [in Russian].

Vinogradov, V.I. (1980) *The Role of the Sedimentary Cycle in Geochemistry of Sulfur Isotopes*. Nauka, Moscow, 192 p. [in Russian].

Vinokurov, A.F., Novikov, Yu.N. & Usatov, A.V. (1997) Fullerenes in geochemistry of endogenous processes. *Geochemistry*, (9), 937–944 [in Russian].

Walt, W.Yu. (1961) Isotopic fractionation of sulfur in geochemical processes. In: *Geochemical Investigations*. Transl. from English. IL, Moscow, pp. 308–332 [in Russian].

Zadonina, N.V., Levi, K.G. & Yazev, S.A. (2007) *Space Hazards in the Geological and Historical Past of Earth: The Time Series Analysis*. IZK SB RAN, Irkutsk, 77 p. [in Russian].

Zavarzin, G.A. (2003a) Evolution of microbial communities. Presentation at the *theoretical seminar on Origin of Living Systems, Denisov Cave Hospital*, August 15–20, 2003 Available from: http:www.bionet.nsc.ru/live/php?f=doclad&p=zavarzin.

Zavarzin, G.A. (2003b) *Microbiology in Natural History: A Course of Lectures*. Nauka, Moscow, 348 p. [in Russian].

Index

T - #0237 - 101024 - C0 - 216/138/5 [7] - CB - 9781138325258 - Gloss Lamination